olids and fluids

The field of matter transport is central to understanding the processing of materials and their subsequent properties. While thermodynamics determines the final stable state towards which a material system tends, it is the kinetics of mass transport that governs how fast it gets there. This textbook gives a solid grounding in the principles of matter transport and their application to a range of materials science and engineering problems.

The author develops a unified phenomenological treatment of this subject applicable to both solids and liquids. Traditionally, mass transport in fluids is considered as an extension of heat transfer theory and uses similar methods and nomenclature. The subject can therefore appear to the student as having little relationship to diffusion in solids. By contrast, the unified approach of this text enables the student to clearly make the connection between these two important fields. The material presented assumes the student has some knowledge of thermodynamics, including the use of binary phase diagrams at a level usually presented in introductory courses on materials. It is also useful, but not essential, for students to have taken a course in partial differential equations. An appendix deals with solution methods for the diffusion equation.

This book is aimed at students of materials science and engineering and related disciplines such as metallurgy and ceramics. It contains numerous worked examples and unsolved problems. The material is suitable for advanced undergraduate and beginning graduate students, and can be covered in a one-semester course.

DAVID WILKINSON received his undergraduate education in Engineering Science at University of Toronto (1972) where he also completed a Master's degree (1974). He did his Ph.D. at Cambridge University (1978) working under the supervision of Prof. Mike Ashby FRS. The field of research was modelling and mechanism maps for pressure sintering. Following a post-doctoral fellowship at University of Pennsylvania he joined the faculty of McMaster University in 1979. He has held visiting fellowships at the Max Planck Institute für Metallforschung Stuttgart (1985/86), University of California Santa Barbara (1991/92) and Institut National des Sciences Appliqués (INSA) in Lyon (1997). He has served on the boards of both the Metallurgical Society of the Canadian Institute of Mining and Metallurgy (CIM) and the Canadian Ceramic Society. In 1996 he was awarded the prize for best materials paper in the *Canadian Metallurgical Quarterly* and in 2000 the Ross Coffin Purdy Medal of the American Ceramic Society (ACerS). Professor Wilkinson's research focusses on the processing and mechanical behaviour of heterogeneous solids, including metals, ceramics and their composites. He has published over 150 research papers. He is a Fellow of CIM and ACerS.

Cambridge Solid State Science Series

The fountains mingle with the river
And the rivers with the ocean
The winds of heaven mix forever
With a sweet emotion;
Nothing in the world is single;
All things by a law divine
In one spirit mix and mingle.
Percy Bysshe Shelley, *Love's Philosophy*

Mass transport in solids and fluids

David S. Wilkinson
Department of Materials Science and Engineering,
McMaster University

Cover photograph: Transmission electron micrograph images of an Al–4 wt% Cu alloy during *in-situ* annealing within the electron microscope at 320 °C. These two micrographs, taken 15 minutes apart, show the progressive dissolution of platelike particles of $CuAl_2$. The field of each micrograph is about 4 µm wide. (Courtesy of Prof. G. C. Weatherly, McMaster University.)

CAMBRIDGE
UNIVERSITY PRESS

PUBLISHED BY THE PRESS SYNDICATE OF THE UNIVERSITY OF CAMBRIDGE
The Pitt Building, Trumpington Street, Cambridge, United Kingdom

CAMBRIDGE UNIVERSITY PRESS
The Edinburgh Building, Cambridge CB2 2RU, UK
40 West 20th Street, New York, NY 10011-4211, USA
10 Stamford Road, Oakleigh, VIC 3166, Australia
Ruiz de Alarcón 13, 28014 Madrid, Spain
Dock House, The Waterfront, Cape Town 8001,
South Africa

http://www.cambridge.org

First published 2000

Printed in the United Kingdom at the University Press, Cambridge

Typeface *Times*   System *3B2*

A catalogue record for this book is available from the British Library

Library of Congress Cataloguing in Publication data

Wilkinson, David S., 1950–
    Mass transport in solids and fluids / David S. Wilkinson.
        p. cm.
    Includes index.
    ISBN 0 521 62409 6 (hb)
        1. Mass transfer. I. Title.

QC318.M3 W55 2000
530.4'75–dc21   00-027646

ISBN 0 521 62409 6 hardback
ISBN 0 521 62494 0 paperback

Contents

Preface

The field of matter transport is central to understanding the processing and subsequent behaviour of materials. While thermodynamics tells us the state in which a material system would like to exist, the kinetics of matter transport tell us how fast it can get there. The materials which we use in service are often not at equilibrium. This is particularly so in solid materials. Here kinetics are sufficiently slow that materials are almost never in complete thermodynamic equilibrium but attain a stable state which is kinetically limited. However, kinetics are equally important in fluids since the transport of matter in a liquid or gas often limits the rate at which many processes for materials fabrication can proceed. The aim of this book is to give students a solid grounding in the principles of matter transport and their application to the solution of engineering problems.

It is this emphasis on engineering application that distinguishes this text from many other books in this field. In the field of solid-state diffusion, for example, much has been written about the mechanisms of diffusion at the atomic scale. This is a fascinating topic and of great interest to materials scientists. However, only an elementary understanding is required in order to treat many practical diffusion problems. Thus, the emphasis throughout this book is on developing methods for solving problems involving the transport of matter in materials. I have attempted throughout to develop a unified view of mass transport in which the approach is independent of the state (whether solid, liquid or gas) of the transport medium. In particular, I have drawn links between the concept of counter diffusion, used to treat problems involving solid-state diffusion in concentrated alloys, and the concept of convective flow in fluids. Both result in a drift velocity which can be handled phenomenologically using the same framework.

This book is aimed at undergraduate students taking degrees in the various disciplines which come under the general umbrella of materials science and engineering (including metallurgy and ceramics). The subject of this book has been for many years treated in a one-term course at McMaster University. It will be suitable for most students in their junior or senior years. The presentation assumes that the students have some knowledge of thermodynamics, and especially in the use of binary phase diagrams, at a level usually presented in introductory courses on materials. It is also useful, but not essential, for students to have done a course in partial differential equations. (An appendix which deals with solution methods for the diffusion equation is offered, primarily as a refresher for those familiar with the subject).

I would like to acknowledge the encouragement and assistance of many colleagues and friends in the preparation of this book. The course on which this book is based was first developed by Prof. Bob Piercy and many of his ideas are incorporated herein. I gratefully acknowledge the contributions of Prof. Gordon Irons in the area of convective mass transfer and that of Prof. Gary Purdy in multicomponent diffusion. I am grateful to both of these colleagues and to a host of students who have commented on early drafts of the book. My thanks go also to Ginnie, Laura and Jan in our Faculty Docucentre for persevering in the preparation of the text, and to Ed McCaffery and John Nychka for the production of many of the figures.

Above all, I wish to express my heartfelt gratitude to Linda, Lisa and Craig for their patience, love and understanding.

David S. Wilkinson
Dundas, Ontario

Principal symbols

Symbol	Definition	Unit
a	lattice parameter	m
a	thermodynamic activity	–
A	cross-sectional area of diffusion path	m^2
C	concentration	mol/m^3
D	diffusion coefficient	m^2/s
D_v, D_a, D_s, D_i	diffusion coefficients for (respectively) vacancies, atoms (self-diffusion), substitutional solute and interstitial solute	m^2/s
D_{0v}, D_{0a}, D_{0s}, D_{0i}	pre-exponential term in Arrhenius expression for the diffusion coefficient	m^2/s
f_i	weight fraction of phase i	–
Gr'	mass transport Grashof number	–
ΔG	free energy	kJ/mol
ΔG_v^f	free energy of vacancy formation	kJ/mol
ΔG_v^m	free energy of vacancy migration	kJ/mol
ΔG_i^m	free energy of interstitial migration	kJ/mol

Symbol	Definition	Unit
j	diffusive flux	$kg/m^2\,s$
J	diffusive flux	$mol/m^2\,s$
k	Boltzmann's constant	$J/atom\,K$
k_D	mass transfer coefficient	m/s
K	equilibrium constant	–
K^*	solubility of gas in solid at 1 atm. pressure	mol/m^3
$\bar{L}$	nominal diffusion distance	m
n	total flux	$kg/m^2\,s$
N	total flux	$mol/m^2\,s$
N_0	Avogadro's number	$/mol$
p	pressure	Pa
p	Boltzmann probability	–
q	flow rate	mol/s or kg/s
Q	activation energy	kJ/mol
R	universal gas constant	$J/mol\,K$
Re	Reynolds number	–
$\dot{r}$	reaction rate	$mol/m^3\,s$ or $kg/m^3\,s$
Sc	Schmidt number	–
Sh	Sherwood number	–
t	time	s
t_e	exposure time	s
T	temperature	K
T_m	melting temperature	K
v	average molar velocity	m/s
v^*	average mass velocity	m/s
V	volume	m^3
X	molar fraction	–
X^*	weight fraction	–
X_v	vacancy concentration	–
X_v^o	equilibrium vacancy concentration	–
$(\Delta X)_{ln}$	log mean driving force (see eq. (7.48))	–
y	position	m
y_I	interface position	m
δ	membrane or layer thickness	m
δ_b	grain boundary thickness	m
η	viscosity	$Pa\,s$
γ	activity coefficient	–
γ^o	Henrian activity coefficient	–
Γ	jump frequency	s^{-1}
λ	mean free path	m

Symbol	Definition	Unit
ν	lattice vibration frequency	s^{-1}
ν	kinematic viscosity	m^2/s
Π	permeability	$m^3(STP)/m\,s\,atm^{1/2}$
ρ	density (mass concentration)	kg/m^3
Ω	atomic or molecular volume	m^3

Part A

Overview

Chapter 1

Introduction to mass transport mechanisms

O world invisible, we view thee,
O world intangible, we touch thee,
O world unknowable, we know thee.
Francis Thompson, *The Kingdom of God*

The transport of matter within materials can occur either by diffusion or by convective flow. Diffusion can occur in both solids and fluids while convective flow is found only in fluids. This chapter provides a brief overview of these processes. It also offers a summary of the mechanisms involved, for diffusion in solids, liquids and gases. While these are presented in a highly simplified fashion they do offer sufficient insight to enable many mass transport problems of practical interest to be solved.

1.1   Mass transport processes

When a drop of dye is added to a beaker of still water the highly concentrated dye spreads throughout the liquid until a uniform pale colour results. There are two processes which can contribute to this. The first is called *diffusion*. This process is driven by differences in the concentration of a substance (in this case the molecules that make up the dye) from one region to another. Diffusion occurs until the concentration becomes uniform, *i.e.* the concentration gradient goes to zero everywhere. The same process happens in the solid state when two soluble substances are mixed together. In this case, however, the process is usually much slower and equilibrium may not be reached in sensible periods of time, especially at low temperatures.

A second process can also contribute to this mixing, at least in fluids (*i.e.* gases and liquids). This involves the bulk motion of the fluid, commonly called *convection*. Convection can carry concentrated packets of the dye into the far reaches of the beaker, thus enhancing the diffusion process. For example, you know intuitively that if you stir the water after adding the dye, the dye will mix much more quickly. Stirring, which induces convection in the liquid, forces regions of high and low dye concentration to mix together, thus reducing the concentration gradients.

1.2   Statement of Fick's First Law

While diffusion can occur in all three *states* (gases, liquids and solids), the mechanisms involved in each at the atomic or molecular scale are likely to be quite different. In a gas, for example, individual molecules are well separated and the rate of diffusion can be understood in terms of the kinetic theory of gases. At the other extreme, we now know that solid-state diffusion is due to the movement of point defects such as vacancies and interstitial atoms within crystalline solids. At the macroscopic scale however, the phenomenon of diffusion is similar whatever state of matter we consider. In fact, this phenomenon was well known and its kinetics could be predicted with reasonable accuracy long before the atomistic mechanisms were properly understood. We can therefore study the process of diffusion without considering these mechanisms in detail, and indeed for most of this book we will largely ignore them. This is not to say that understanding diffusion mechanisms is not important. There are a range of important phenomena which cannot be understood otherwise, and we will return to this important topic later on. But, for the most part, we will consider a range of engineering problems involving diffusion for which a phenomenological understanding of the diffusion process is sufficient.

What do I mean by a 'phenomenological' understanding? This is simply another way of saying that we will use empirical relationships which describe the 'phenomenon' of diffusion as the basis for solving problems. The fundamental relationship in this case is due to Adolf Fick and was first developed by him in 1855. Fick surmised, on the basis of experiment and intuition, that diffusion occurs in solids in response to a concentration gradient. The process was thought to be analogous to that of heat transfer which had been analyzed by Fourier some years previously. Suppose we consider a binary mixture of two materials, say B dissolved in A. Wherever a concentration gradient exists (as illustrated in Fig. 1.1) Fick suggested that diffusion should occur in order to reduce the gradient. This produces a 'flux', defined as the rate of flow per unit of cross-sectional area.[¶] Once the concentration gradient is removed then the flux should be zero. The simplest possible equation that relates a flux to the concentration gradient which drives it, and that meets these conditions, is

$$J_B = -D_B \frac{\partial C_B}{\partial y}. \tag{1.1}$$

Here, J_B is the flux of solute B in the A–B solution, while C_B is the concentration of this solute. The ∂ symbol represents a partial derivative. The equation is clearly written for a concentration gradient in one dimension only, along the y-axis (as illustrated in Fig. 1.1). The minus sign is required since the flux drives solute 'down' a concentration gradient. Finally, we have introduced a proportionality constant D_B which we call the 'diffusion coefficient'. This equation is generally referred to as *Fick's First Law*.

Of course we can always write an equation of this sort but it is not very useful unless D_B is independent of position y, or least approximately so. This really means that D_B should be independent of the solute concentration. As we will see, while this is not universally the case it is usually so to an adequate level of approximation, enabling many problems to be solved using this simple governing equation.

Before proceeding we should investigate the units of the various parameters in Fick's First Law. We generally use units of moles per unit volume (*e.g.* mol/m³) for concentration, although concentration can also be represented in terms of mass per unit volume (*e.g.* kg/m³). The flux has units of matter (either in moles or mass) per unit of cross-sectional area per unit time (*i.e.* mol/m² s or kg/m² s).

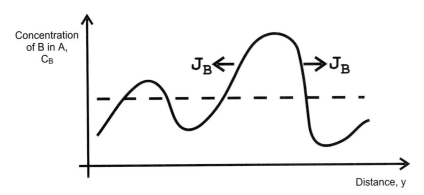

Figure 1.1 A schematic illustration of the diffusive response to a concentration gradient.

[¶] The rate of matter flow will depend on the width of the path through which diffusion occurs, *i.e.* for a given concentration gradient we expect twice as much flow if the piece has an area of 2 cm² than if it had an area of only 1 cm². (This is the same as saying that for cars travelling at 100 km/h and spaced 100 m apart the rate of cars passing a given point is twice as high on a four-lane road than on a two-lane road). By normalizing flow rates by the cross-sectional area of the path to define the 'flux' we remove this dependence from consideration.

The units of the diffusion coefficient can then be deduced. They are, in SI units, given as m^2/s. You should note that whichever unit of concentration you choose (*i.e.* based on either mass or moles) the units of flux you use must be consistent.

1.3   Mechanisms of diffusion

While many diffusion problems can indeed be solved without a detailed understanding of the mechanisms involved, it is useful to develop some physical picture of the process involved at the outset. We will discuss this here only in general terms. We want merely to develop a conceptual picture of a solid, liquid or gas and the way in which atoms are able to move around within them.

In the *condensed* states (*i.e.* solids and liquids), diffusion is known to occur due to a random process in which atoms or molecules are able to exchange positions with neighbours due to random thermal fluctuations. Therefore diffusion belongs to a class of basic physical processes we refer to as *thermally activated processes*. Like all such processes the kinetics are temperature dependent, and obey an Arrhenius equation of the form

$$\text{rate} \sim \exp\left(-\frac{Q}{RT}\right),$$ (1.2)

where T is the absolute temperature and R is the universal gas constant ($8.314\,J/mol\,K$). The parameter Q is the activation energy for the process, having units of energy per mole and is related to the thermal energy (or, to be more precise, the enthalpy) needed for an atom or molecule to overcome the barrier between one stable position and another. This equation leads to a substantial temperature dependence. For example, with an activation energy of $200\,kJ/mol$, typical for many diffusion processes in solids, the rate increases by almost a factor of 15 following a 100 deg K increase in temperature from 900 to 1000 K. Moreover, at room temperature a similar 100 deg K temperature increase would increase the rate by over nine orders of magnitude. Therefore, processes that proceed at a reasonable rate at high temperature do not seem to occur at all if we cool things down. Examples of this are readily seen around us. Consider a plate-glass window. This has the same structure at $800\,°C$ and room temperature. However at the high temperature it flows readily and can be easily shaped. At room temperature it holds its shape without any visible change for years. This is because the viscous flow of a glass is a thermally activated process.

Example 1.1: Effect of temperature on diffusion in liquids

In a typical liquid the activation energy for diffusion is $10\,kJ/mol$. Estimate by how much the diffusion coefficient is increased on raising the temperature from T_1 to T_2, say 20 to $120\,°C$.

According to eq. (1.2) we can write the ratio of diffusion rates at the two temperatures as

$$\frac{\text{rate}_{T_2}}{\text{rate}_{T_1}} = \exp\left[\frac{Q}{R}\left(\frac{1}{T_1} - \frac{1}{T_2}\right)\right].$$

By substitution into this equation we find that the ratio is

$$\frac{\text{rate}_{T_2}}{\text{rate}_{T_1}} = \exp\left[\frac{10,000}{8.314}\left(\frac{1}{293} - \frac{1}{393}\right)\right]$$

which is equal to 2.84. In other words the small activation energy found in liquids leads to a much weaker temperature dependence than that for solid-state diffusion.

1.4 Diffusion in solids

In crystalline solids diffusion occurs by the movement of *point defects*. There are several different types of point defects, but we will be primarily interested in just two – vacancies and interstitial atoms. A vacancy is simply an unoccupied lattice site, as shown in Fig. 1.2.

These defects are created by thermal fluctuations within a lattice and any crystal has an equilibrium vacancy concentration X_v^o which increases with temperature according to an Arrhenius relationship of the form

$$X_v^o = \exp\left(-\frac{\Delta G_v^f}{kT}\right), \tag{1.3}$$

vacancy

substitutional solute

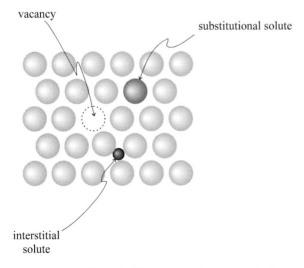

interstitial
solute

Figure 1.2 A schematic illustration of a vacant site in a crystalline solid, called a 'vacancy', and solute atoms located at both interstitial and substitutional sites.

where ΔG_v^f is the free energy of formation of a vacancy, and k is Boltzmann's constant.[¶] These vacant lattice sites play a major role in diffusion, as we will see.

Example 1.2: Vacancy concentration in a typical metal

The free energy of vacancy formation in aluminum is 0.76 eV per vacancy. Estimate the vacancy concentration in this material just below the melting temperature (660 °C).

Free energies for atomic processes are often listed in units of electron volts (eV). The conversion to SI units is: 1 eV per atom = 96.5 kJ/mol. Therefore the free energy of vacancy formation is 73.3 kJ/mol. We now substitute this into eq. (1.3) to find the equilibrium vacancy concentration at 660 °C = 933 K. The answer is 7.9×10^{-5}. This is a typical value for a metal near its melting point, *i.e.* $X_v^o \approx 10^{-4}$.

Foreign or solute atoms can enter a crystal either by occupying the same lattice sites as the host atom (in which case we refer to the site as 'substitutional') or else they can occupy sites between those occupied by the host atoms (in which case we refer to the site as 'interstitial'). Both cases are illustrated in Fig. 1.2.

1.4.1   Interstitial diffusion

We start with the diffusion of interstitial atoms since this is the simplest to understand. Solute atoms which are much smaller than the host enter the lattice interstitially (see Fig. 1.2). The solubility of such atoms is, however, usually rather limited. Therefore almost all interstitial atoms are well-separated from one another. From a diffusional viewpoint this means that all of the neighbouring interstitial sites around a particular solute atom are usually vacant. These sites are therefore available for the solute atom to jump to. However, in order for an atom to jump to one of these sites from its current position it must squeeze through the narrow channel that separates stable sites. This requires thermal energy, also referred to as vibrational energy. The change of free energy of an atom with position is illustrated schematically in Fig. 1.3. The activation barrier, which an atom (or ion) must overcome to move from one stable position to another, has a height ΔG_i^m, where the subscript and superscript refer to 'interstitial' and 'migration' respectively. Whether a given interstitial atom has sufficient vibrational energy to overcome this barrier is determined by Boltzmann statistics which describe how the thermal energy of a large assemblage of atoms is distributed. While the average energy of the atoms is proportional to kT, at any given instant some atoms

---

[¶] Note that the universal gas constant R, and Boltzmann's constant k, are related by Avogadro's number N_0, such that $R = k \cdot N_0$. We can use these interchangeably in an Arrhenius relationship, providing that we adjust the activation energy, *i.e.* with k the energy per atom is required while with R the energy per mole is used.

will have higher than average energy while others have less. The probability that any given atom has an excess thermal energy greater than ΔG_i^m is given by

$$p = \exp\left(-\frac{\Delta G_i^m}{kT}\right). \tag{1.4}$$

Because of the exponential dependence, this probability increases steeply with temperature, which explains why the rates of diffusion-controlled processes are highly temperature sensitive. Atoms vibrate back and forth within each site with a characteristic frequency ν. This is known as the Debye frequency and, from the theory of harmonic oscillators, it can be shown to be proportional to (atomic mass)$^{-1/2}$, and of order $10^{12}\,\text{s}^{-1}$. The frequency Γ, with which an interstitial atom jumps from its current position to an adjacent site, is just the product of the vibration frequency ν and the probability of making a successful jump p. Thus

$$\Gamma = \nu p \tag{1.5}$$

Now in order to relate this to the process of diffusion we need to think about how these atomic jumps, multiplied many times over, affect an initially non-uniform distribution of solute atoms. Let us consider a simple one-dimensional problem as illustrated in Fig. 1.4. Two adjacent lattice planes are separated by a distance a, the lattice spacing. We will call these two planes 1 and 2. There are n_1 solute atoms in plane 1 and n_2 in plane 2. Atoms in plane 1 can jump either right or left. We will assume that they do so with equal likelihood. Therefore they have a jump frequency $\frac{1}{2}\Gamma$ in each direction. Thus the number of solute atoms jumping from plane 1 to plane 2 in a time interval δt is $\frac{1}{2}\Gamma \cdot n_1 \cdot \delta t$, while the number of solute atoms jumping from plane 2 to plane 1 in a time interval δt is $\frac{1}{2}\Gamma \cdot n_2 \cdot \delta t$. The difference between these two

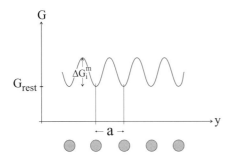

Figure 1.3 When small atoms enter a crystal, they occupy positions between the lattice atoms, called interstitial sites. The free energy of an atom depends on its position. The points of lowest energy are the equilibrium, or rest, positions with energy G_{rest}. In order to move between these positions, atoms must overcome an energy barrier. For interstitial atoms, this barrier has height ΔG_i^m.

values produces a net flow of solute. If we divide the net flow by the cross-sectional area of the plane A and by the time interval δt, this becomes a net flux

$$J = \frac{1}{2}\frac{\Gamma}{A}(n_1 - n_2).$$

The concentration C of solute within each plane is given by $n/(Aa)$. Therefore the flux can be written as

$$J = \tfrac{1}{2}\Gamma a(C_1 - C_2). \tag{1.6}$$

We now want to convert the difference in the concentration of solute atoms in each plane to a concentration gradient

$$J = -\tfrac{1}{2}\,\Gamma a^2\,\frac{\partial C}{\partial y}.$$

This is, of course, just Fick's First Law with the diffusion coefficient now given by $D = \tfrac{1}{2}\Gamma \cdot a^2$. Note that this equation does not depend on the mechanism by which diffusion occurs. It is therefore a general result. We have therefore shown that the empirical equation first developed by Fick also has a sound basis in theory.

It is important to recognize that this model treats diffusion as a purely random process. The individual atoms have no sense of the concentration gradient, *i.e.* they do not feel any force driving them towards regions of lower concentration.

Before proceeding we should generalize this result to three dimensions. This is done simply by replacing the $\tfrac{1}{2}$ term by a more general geometric parameter g, which represents the larger range of possible directions in which an atom can jump. Generally, $g \approx 1/6$. Thus

$$J = -g\Gamma a^2\,\frac{\partial C}{\partial y}. \tag{1.7}$$

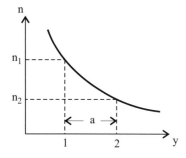

Figure 1.4 A schematic illustration of a concentration profile. The difference in jump rate from plane 1 to plane 2, as compared with the rate from plane 2 to plane 1, results in a net flux.

The mechanism by which diffusion occurs affects the result through the jump frequency term. For the interstitial case we substitute eqs. (1.4) and (1.5) into eq. (1.7). Therefore the interstitial diffusion coefficient is given by

$$D_i = g\Gamma_i \cdot a^2 = g\nu a^2 \exp\left(-\frac{\Delta G_i^m}{kT}\right) \tag{1.8}$$

where the subscript 'i' is used throughout to indicate that these terms refer specifically to the interstitial mechanism of diffusion. We can determine the temperature dependence of the diffusion coefficient more precisely by separating the free energy into two terms for the enthalpy and entropy, such that $\Delta G_i^m = \Delta H_i^m - T \cdot \Delta S_i^m$. By substituting this into eq. (1.8) we get

$$D_i = g\nu a^2 \exp\left(\frac{\Delta S_i^m}{k}\right) \exp\left(-\frac{\Delta H_i^m}{kT}\right) = D_{0i} \exp\left(-\frac{Q_i}{kT}\right). \tag{1.9}$$

Here all of the temperature-independent terms are combined into a single parameter D_{0i}, while the enthalpy has been rewritten as an activation energy Q_i. These two parameters are all that is required to characterize the diffusion coefficient and they are found tabulated in many compilations of diffusion data, such as that given in Appendix B.

1.4.2 Vacancy diffusion

When large atoms[¶] enter a lattice they do so 'substitutionally', *i.e.* they use the same lattice positions as the host atoms. It is difficult for these solute atoms, and for the host atoms themselves, to diffuse through the lattice. In a perfect lattice, with every site occupied by a host or substitutional atom, there would be no sites to which atoms could jump. However as we have already discussed, each lattice contains a certain fraction of vacant sites, *i.e.* vacancies. Whenever an atom is sitting next to a vacancy it can move into that unoccupied site. Otherwise it is immobilized. We therefore have to take this into account in calculating the jump frequency of host atoms and substitutional solutes. We will start by considering the diffusion of vacancies themselves.

The concentration of vacancies in crystalline solids varies considerably with temperature. However, vacancies are always present in dilute concentrations. (For example, as we noted in Example 1.2, the vacancy concentration in a face-centred-cubic metal at its melting point is about 10^{-4} and decreases rapidly as the temperature drops). Therefore, vacancies are inevitably surrounded by occupied lattice sites. Just as interstitial atoms must overcome an activation barrier in order to jump between adjacent sites, so must vacancies. Thus the probability that a vacancy has sufficient energy ΔG_v^m to move

---

[¶] In this case 'large' means atoms whose radii are comparable with or larger than that of the host atoms.

(or conversely that a host atom has sufficient energy to jump into an adjacent vacant site) is still given by the Boltzmann distribution but with a different height to the activation barrier. Thus

$$p = \exp\left(-\frac{\Delta G_v^m}{kT}\right). \tag{1.10}$$

The jump frequency for a vacancy moving to a specific adjacent site is therefore still given by eq. (1.5), but with the probability term given by eq. (1.10). The diffusion coefficient of vacancies is therefore equal to

$$D_v = g\Gamma_v \cdot a^2 = g\nu a^2 \exp\left(-\frac{\Delta G_v^m}{kT}\right), \tag{1.11}$$

where the parameter 'a' now represents the nearest neighbour spacing in the lattice.

While the diffusion coefficient for vacant sites is of interest, what we really want is the diffusion coefficient of lattice atoms – either host atoms or a substitutional solute. This is much smaller than that for vacancies since most lattice atoms are not adjacent to vacant sites and therefore cannot move. For lattice atoms the jump frequency must be altered by including the probability that the atom is adjacent to a vacant site. Thus

$$\Gamma_a = X_v \cdot \Gamma_v \tag{1.12}$$

where the subscript 'a' refers to lattice atoms and the 'v' to vacancies. If we assume that the vacancy concentration is at equilibrium then X_v is given by eq. (1.3).[¶] Therefore the lattice diffusion coefficient of a pure material (generally referred to as the 'self-diffusion' coefficient) is given by

$$D_a = g\Gamma_a \cdot a^2 = g\nu a^2 \exp\left(-\frac{\Delta G_v^f + \Delta G_v^m}{kT}\right). \tag{1.13}$$

A very similar expression follows for the diffusion of a substitutional solute except that the free energy of motion for the solute ΔG_s^m may be somewhat different from that of the host atom, *i.e.*

$$D_s = g\nu a^2 \exp\left(-\frac{\Delta G_v^f + \Delta G_s^m}{kT}\right). \tag{1.14}$$

A more complete analysis would incorporate effects due to the crystal structure of the material. From this we would learn that g is a function of crystal symmetry.

[¶] This is generally a good assumption. For the reasons just discussed the diffusion coefficient of vacancies is orders of magnitude higher than that for atoms. Thus following a temperature change (whether up or down), the vacancy concentration will re-equilibrate rather quickly compared with the time required for other diffusion processes to occur.

Example 1.3: Relative magnitude of interstitial and self-diffusion

Given the data in Appendix B, determine the ratio of diffusion coefficients for carbon and iron in a carbon steel at 1000 °C.

At this temperature iron exists in the form of austenite, with carbon as an interstitial solute. After consulting the data we find that the pre-exponential terms for C and Fe diffusion in γ-Fe are quite similar ($1.5 \times 10^{-5} \, \text{m}^2/\text{s}$ for carbon interstitials and $2.2 \times 10^{-5} \, \text{m}^2/\text{s}$ for self-diffusion). However, the activation energy for C in Fe is much lower than that for self-diffusion (142 vs 268 kJ/mol). Thus the ratio of the diffusion coefficients is

$$\frac{D_\text{C}}{D_\text{Fe}} = \frac{1.5 \times 10^{-5}}{2.2 \times 10^{-5}} \exp\left(\frac{268{,}000 - 142{,}000}{8.314 \times 1273}\right) = 1.01 \times 10^5$$

i.e. the rate of interstitial diffusion (of carbon) exceeds that for self-diffusion (of iron) by five orders of magnitude, at this temperature.

How will this ratio vary with temperature?

The activation energy is greater for the diffusion of iron than for the interstitial carbon. Therefore the diffusion coefficient of iron increases more rapidly as the temperature rises. This means that the ratio between these two diffusion coefficients will fall as the temperature increases and will rise as the temperature goes down.

1.4.3   Diffusion in alloys

As we proceed through an in-depth discussion of diffusion within alloy systems we will encounter a range of diffusion coefficients other than those just discussed. The diffusion of vacancies and of interstitial atoms are most easily dealt with because these species are always in dilute solution. Similarly the diffusion coefficient of a *dilute* substitutional solute is given by eq. (1.14). The self-diffusion coefficient is often measured using radioactive tracers, *i.e.* a *dilute* concentration of a radioactive isotope of the same substance is dissolved into the lattice and its rate of diffusion is followed by measuring the spread of radioactivity within the solid. Once again the solute (in this case the radioactive isotope) is dilute and eq. (1.14) is applicable.

Diffusion within substitutional alloys is complicated at higher solute concentrations by the interaction between the elements. In an A–B substitutional solid solution, since the number of lattice sites must be conserved, B atoms can only diffuse in a given direction if there is a counter diffusion of A atoms in the opposite direction. How then are the diffusion coefficients of the two (or more) species related? An analysis of this problem (which we shall discuss in detail in Chapter 7) leads to an *interdiffusion* coefficient for binary alloys of the form

$$\tilde{D} = X_\text{A} D_\text{B} + X_\text{B} D_\text{A}, \tag{1.15}$$

where X_A and X_B are the mole fractions of A and B, while D_A and D_B are the diffusion coefficients of the individual species in the alloy. It is this diffusion coefficient which is substituted into Fick's First Law in order to determine the flux induced by a concentration gradient. Note, however, that for a dilute solution $\tilde{D}$ is dominated by the diffusion coefficient of the dilute species. This is consistent with all of the preceding discussion regarding dilute alloys, and the equation for diffusivity of a dilute solute can be considered to be a special case of the interdiffusion coefficient.

1.4.4   Diffusion in compounds

In *ordered* compounds such as intermetallic materials or ionic solids, an additional level of complexity is introduced. In general, the structure consists of two or more interlocking sublattices and each individual species is confined to its own sublattice. In Al_2O_3 for example the Al^{3+} and O^{2-} ions each have a unique set of lattice positions. Therefore an Al^{3+} ion will not jump into an unoccupied O^{2-} site even if it is an immediate neighbour. There is a very strong electrostatic repulsion in this case which prevents such mixing. The vacancy concentration on each sublattice is also different. Diffusion therefore occurs separately on each sublattice and we can assign a separate diffusion coefficient to each species. However, just because diffusion occurs on separate sublattices does not mean they are independent. In order to maintain the correct ratio of species in a compound, diffusion will be linked, with the slower diffusing species controlling the overall rate. We will consider examples of this type of linked diffusion throughout the text. However, a detailed discussion of such phenomena is best left to advanced study.[¶]

1.5   **Diffusion in liquids**

The more random distribution of molecules and their greater mobility makes diffusion in liquids considerably easier than in solids, *i.e.* the diffusion coefficients should be considerably larger. Moreover, the random structure should reduce the height of the activation barrier between stable low-energy positions. Thus we expect diffusion in liquids to exhibit low activation energies, and thus be relatively insensitive to temperature. However, the development of a theory for molecular motion (either viscous flow or diffusion) is hampered by our poor understanding of the structure of liquids. It is known that liquids possess some short-range order on the scale of the molecular spacing, but that they are disordered beyond this. Molecular motion therefore involves a local re-arrangement process whereby the clusters can change their size and orientation,

---

[¶] See, for example, Y.-M. Chiang, D. Birnie, W. D. Kingery, *Physical Ceramics: Principles of Ceramic Science and Engineering* (Wiley, New York, 1997)

and molecules can transfer from one cluster to another. These processes are thermally activated so that the viscosity of a liquid exhibits Arrhenius behaviour, *i.e.*

$$\eta = \eta_0 \exp\left(+\frac{\Delta G_{\mathrm{vis}}}{kT}\right). \tag{1.16}$$

Here ΔG_{vis} is the free activation barrier for molecules to rearrange within the liquid. There is no first-principles model which adequately predicts the value of these parameters, although a theory by Eyring[¶] suggests that

$$\eta_0 \approx \frac{N_0 h}{\bar{V}}, \tag{1.17}$$

where N_0 is Avogadro's number, h is Planck's constant, and $\bar{V}$ is the molar volume.

Most models for diffusion in liquids relate this process directly to viscous flow, and lead to equations for diffusivity as a function of viscosity. One of the earliest, simplest, and surprisingly most successful was developed by Einstein.[†] It considers diffusing molecules as hard spheres moving through a viscous continuum according to Stokes law. The result is that

$$D = \frac{kT}{4\pi r \eta} \tag{1.18}$$

where r is the molecular (*i.e.* hard-sphere) radius. Combining this with eq. (1.16), we see that the diffusion equation again has the form $D = D_0 \exp(-Q/RT)$ where $Q = \Delta H_{\mathrm{vis}}$, the activation energy (or enthalpy) for viscous flow and

$$D_0 = \frac{kT}{4\pi r \eta_0} \exp\left(-\frac{\Delta S_{\mathrm{vis}}}{r}\right). \tag{1.19}$$

Although the activation energy is not easily calculated from models, it is clear that molecular motion in liquids will involve smaller barriers than for solids. Therefore activation energies will be smaller, and the temperature dependence of the diffusion coefficient can be expected to be considerably less in liquids than in solids.

1.6   Diffusion in gases

In condensed matter, atomic or molecular motion occurs in small jumps of atomic dimensions. In a gas, however, molecules can travel large distances before being scattered by collisions with other molecules. The theoretical

[¶] H. Eyring, S. Glasstone and K. Laidler, *Theory of Rate Processes* (McGraw-Hill, New York, 1941)
[†] A. Einstein, *Ann. Phys. (Leipzig)*, **17**, 549 (1905)

basis for this, called the *kinetic theory of gases*, is now well established. In this model, atoms are treated as hard spheres of fixed diameter a, and mass m. In a binary system, the theory is made more complicated by differences in the size and mass for each species. We therefore illustrate the theory using the inter-diffusion of two species B and B* having equal mass and size. Kinetic theory suggests that the mean velocity of a gas, $\bar{v}$ is given by

$$\bar{v} = \sqrt{\frac{8kT}{\pi m}}, \qquad (1.20)$$

where m is the particle mass, while the mean free path between collisions is

$$\lambda = \frac{1}{\sqrt{2}\pi a^2 C}, \qquad (1.21)$$

where C is the total gas concentration (molecules/unit volume).

We need to express the velocity in terms of a flux J (molecules per unit area per unit time). Since this is a general result (equally applicable in solids and liquids as well as gases) and we will often have use of it, let us take the time to derive it here. Suppose that the gas is enclosed in a narrow cylindrical tube of radius R and length L. We therefore need only consider the axial velocity $\bar{v}_z$. Consider a small slice of length dz. The number of B* molecules in this slice is equal to $dn = \pi R^2 dz \cdot C_{B^*}$. It takes a length of time equal to $dt = dz/v_z$ for the molecule to pass through this slice. But this produces a flux equal to $dn/(\pi R^2 dt) = C_{B^*} v_z$, i.e. the flux J in a given direction produced by a drift velocity v in the same direction is

$$J = C_{B^*} \cdot v. \qquad (1.22)$$

This result can easily be generalized to three dimensions by recognizing that molecules can travel in three orthogonal directions, with a plus or minus orientation. Thus, the flux in any given direction due to a random velocity of average magnitude given by $\bar{v}$, is

$$J = \tfrac{1}{3} C_{B^*} \bar{v}. \qquad (1.23)$$

We can determine the net flux due to a concentration gradient in a similar manner to the approach we used for solids. However, we now consider two parallel planes separated by a distance λ, at say $z = z_0$ and $z = z_0 + \lambda$. The flow rate per unit area for gas going from z_0 to $(z_0 + \lambda)$ is $\tfrac{1}{3} C_{B^*}(z_0)\bar{v}$, while the flow rate per unit area for gas going from $(z_0 + \lambda)$ to z_0 is $-\tfrac{1}{3} C_{B^*}(z_0 + \lambda)\bar{v}$.

Therefore, the net flux is

$$J_{B^*} = \tfrac{1}{3}\bar{v}[C_{B^*}(z_0) - C_{B^*}(z_0 + \lambda)] = -\tfrac{1}{3}\bar{v}\lambda \frac{\partial C_{B^*}}{\partial z}. \qquad (1.24)$$

Thus the diffusion coefficient is

$$D = \tfrac{1}{3}\bar{v}\lambda$$

$$= \frac{2}{3\pi a^2 C}\sqrt{\frac{kT}{\pi m}}. \tag{1.25}$$

For an ideal gas we can express the concentration as a pressure, $C = p/kT$. We can therefore rewrite the diffusion coefficient as

$$D = \frac{2}{3a^2 p}\sqrt{\frac{(kT)^3}{\pi^3 m}}. \tag{1.26}$$

Example 1.4: Diffusion coefficient for oxygen in air

Use eq. (1.26) to estimate the diffusion coefficient of oxygen in 1 atmosphere of air at room temperature. Note that while this equation is not strictly valid for a gas mixture, since nitrogen and oxygen have similar size and molecular weight this equation gives a reasonable estimate.

The partial pressure of oxygen in 1 atm. of air is 0.21 atm or 2.1×10^4 Pa. Oxygen is a diatomic molecule with a mass of 32 g per mole or 5.23×10^{-23} g/molecule. It has a diameter of 0.26 nm. At room temperature kT is equal to 4.1×10^{-21} J/molecule. On substituting these parameters into eq. (1.26) we find that the diffusion coefficient of oxygen is 3.06×10^{-6} m^2/s.

If we have two gases A and B with unequal mass and diameter, the result is similar but more complex in form. It is

$$D = \frac{2}{3p}\left(\frac{kT}{\pi}\right)^{3/2}\sqrt{\left(\frac{1}{2m_A}+\frac{1}{2m_B}\right)}\frac{1}{(\tfrac{1}{2}a_A + \tfrac{1}{2}a_B)^2} \tag{1.27}$$

Kirkaldy and Young[¶] present data which shows the predictions from this theory are quite accurate for a wide range of gas mixtures.

1.7   Diffusion data

There is a large amount of diffusion data available in the literature. Unfortunately, it is rather widely scattered and finding the exact data you need is not always easy. Most texts, including this one, contain a limited amount of data, generally for the most commonly used materials. A useful compilation of tracer diffusion data is that given by Askill.[†] Some computerized databases,

[¶] J. S. Kirkaldy and D. J. Young, *Diffusion in the Condensed State*, (Institute of Metals, London, 1987), p.3

[†] John Askill, *Tracer Diffusion Data for Metals, Alloys and Simple Oxides* (IFI/Plenum, New York, 1970)

such as TAPP,[¶] now contain useful compilations as well. Also, there is a journal, *Diffusion Data*, which is dedicated to this subject.

A separate issue is the accuracy of diffusion coefficients. Some data is collected by direct methods such as radioactive tracer measurements in solids. These are generally regarded to be the most accurate. Often, however, diffusion data is inferred from indirect measurements such as oxidation or sintering kinetics. This requires a model which may not always be valid. In addition, diffusion coefficients are affected by the presence of impurities. This is especially so for ionic solids. In these materials the point defect concentrations can change by orders of magnitude as the impurity level changes. There are two things to learn from this. The first is the need to exercise considerable caution in using diffusion data. We must not only ensure that the data is accurate, but also that it has been measured on material that is similar to the material on which we are working (including minor impurities). Second, once the accuracy of the available data has been assessed we should use a solution method of similar accuracy. For example in a given materials system the diffusion data which is available may only be accurate to within a factor of two. So there is no point in developing a sophisticated finite-element computer simulation capable of predicting concentration profiles with an accuracy of three significant figures. A simple analytical solution is all that can be justified.

In many systems no diffusion data is available. It is still possible, however, to make reasonable estimates of diffusivities by extrapolation from other systems. For example, we noted above that the diffusion coefficient of a dilute substitutional solute within a give host lattice differs from that of the host atom (and other solutes) by the term for the free energy of migration into a vacancy. This energy is largely determined by the size and valence of the atom. Thus, the diffusion coefficient of a different atom but with similar size and valence in the same host lattice will give a fairly good estimate. Brown and Ashby have developed empirical correlations for the self-diffusion coefficient of materials based on materials class and crystal structure, see Table 1.1. These correlations work best for simple structures such as metals and alkali halides but are less reliable for more complex materials. Table 1.1 provides data on the diffusion coefficient at the melting point $D(T_m)$, and the normalized activation energy Q/RT_m. From this data both Q and D_0 can be inferred since $D_0 = D(T_m) \cdot \exp(Q/RT_m)$.

Example 1.5: Estimated diffusion coefficients

Use the data from Table 1.1 to estimate the diffusion coefficient of Cr at 1000 °C. Compare this with the measured diffusion coefficient of Cr using data from Appendix B.

[¶] *Thermochemical and Physical Properties Database* (E. S. Microware Inc., Hamilton, OH)

Table 1.1 Diffusion coefficient correlations.

Material class	$D(T_m)$ (m^2/s)	Q/RT_m
fcc metals	5.5×10^{-11}	18.4
bcc (group I) metals	1.4×10^{-8}	14.7
bcc transition metals	2.9×10^{-10}	17.8
hcp metals	1.6×10^{-10}	17.3
alkali halides	3.2×10^{-11}	22.7

Source: A. M. Brown and M. F. Ashby, *Acta metall.*, **28**, 1085 (1980)

The melting temperature T_m of Cr is 2403 K. From this we can estimate the activation energy:

$Q = 18.4 \, RT_m = 3.67 \times 10^5 \, \text{J/mol} = 367 \, \text{kJ/mol}.$

To estimate D_0 we use the value of $D(T_m)$ given in the table and, knowing

$$D(T_m) = D_0 \exp\left(-\frac{Q}{RT_m}\right),$$

we calculate $D_0 = 5.5 \times 10^{-11} / \exp(-18.4) = 5.39 \times 10^{-3} \, m^2/s$. Thus the estimated diffusion coefficient at 1000 °C is

$$D = 5.39 \times 10^{-3} \exp\left(-\frac{367,000}{8.314 \times 1273}\right) = 4.70 \times 10^{-18} \, m^2/s.$$

From Appendix B data we estimate the diffusion coefficient to be

$$D = 2.0 \times 10^{-5} \exp\left(-\frac{308,000}{8.314 \times 1273}\right) = 4.60 \times 10^{-18} \, m^2/s.$$

The agreement is remarkably good. You should note however that because of the difference in activation energy the disagreement will be larger at other temperatures.

1.8 Mass transport by convection

At the beginning of this chapter we introduced two processes for mass transport in materials – one was diffusion; the other was convection. This latter process involves the collective movement of molecules. It therefore applies in fluids (*i.e.* liquids and gases) but not solids. During convective flow the fluid moves with an average velocity v, giving rise to a flux of any species within the fluid, given by the formula derived in eq. (1.22).

An analysis of mass transport kinetics by convection therefore requires analyzing the flow characteristics of the fluid. Fluid mechanics is a separate field in its own right and a detailed development of this field is beyond the scope of this book. It is clear however that mass transport under conditions in which convection is possible involves a coupling of fluid flow and diffusion.

The nature of this coupling depends on the nature of the flow processes involved. In many cases for example we are interested in the transfer of matter between a solid surface and a flowing fluid. This transfer takes place in a thin region near the surface, called a boundary layer. If the fluid flow within this region is turbulent then the solute concentration within this region will be constant and mass transfer will be controlled by the reaction rate at the interface. If however, and this is the more common case, the boundary layer involves laminar flow then a concentration gradient will develop between the surface and the bulk fluid. In this case transport will be controlled by diffusion through the boundary layer.

While this introduction gives only the briefest flavour of how convective flow leads to mass transport, it will suffice for now. We will leave any further discussion to a later chapter, after we have completed our analysis of diffusion in the absence of convection.

1.9   Further reading

P. Shewmon, *Diffusion in Solids*, 2nd edition (TMS-AIME, Warrendale, PA, 1989)

D. R. Poirier and G. H. Geiger, *Transport Phenomena in Materials Processing* (TMS-AIME, Warrendale, PA, 1994)

D. R. Gaskell, *An Introduction to Transport Phenomena in Materials Engineering* (Macmillan, New York, 1992)

R. J. Borg and G. J. Dienes, *An Introduction to Solid State Diffusion* (Academic Press, San Diego, CA, 1988)

1.10   Problems to chapter 1

1.1   (a) Starting from first principles use atomic jump theory to show that the self-diffusion coefficient for atoms diffusing by a vacancy mechanism is

$$D = ga^2 \nu \exp\left[-\frac{(\Delta G_v^f + \Delta G_v^m)}{kT} \right]$$

(b) How does the diffusion coefficient depend on temperature, *i.e.* what material parameter(s) control the T-dependence.

(c) By how much does the diffusion coefficient of Al change on going from 200 to 400 °C?

Data: $a = 2 \times 10^{-10}$ m
$\Delta S_v^f = S_v^m = 1.5 \, \text{k/atom}$
$\Delta G_v^f = 0.76 \, \text{eV}$
$\Delta G_v^m = 0.63 \, \text{eV}$

1.2   Given the data below determine by what ratio the self-diffusion coefficient of pure copper will change when the material is heated from 300 to 500 °C.

Data: Vibration frequency, $\nu = 10^{12}/\text{s}$
$\Delta H_v^f = 113 \, \text{kJ/mol}$
$\Delta S_v^f = 1.5R \, \text{kJ/mol K}$
$\Delta H_v^m = 106 \, \text{kJ/mol}$
$\Delta S_v^m \approx 0$
Atomic spacing, $a = 0.25 \, \text{nm}$

Part B

Solid-state diffusion in dilute alloys

In this part of this book we restrict our attention to diffusion in the solid state involving dilute binary alloys. These restrictions are useful for a variety of reasons. In the solid state, convection is not possible and thus all mass transport is by diffusion only. Binary alloys, *i.e.* alloys containing only two components, are easier to treat than ternary and higher systems. Moreover, the thermodynamic information which we will need is neatly summarized in binary phase diagrams which are readily available for most systems of technological interest. Finally, the restriction to dilute alloys simplifies our treatment of the diffusion coefficient, since for dilute alloys diffusion is always controlled by the solute.

The various complexities associated with relaxing these assumptions will be treated in Part C, devoted to mass transport in concentrated alloys and fluids.

Chapter 2

Steady-state diffusion

As aromatic plants bestow
No spicy fragrance while they grow;
But crushed or trodden to the ground,
Diffuse their balmy sweets around.
Oliver Goldsmith, The Captivity, act i.

When external conditions establish different, but fixed, solute concentrations on either side of a solid, steady-state conditions will eventually be established. Problems involving steady state are best solved using Fick's First Law of diffusion. In this chapter we review this Law and extend it to three dimensions. We then consider its application to steady-state problems, including the permeability of gases through a solid membrane and diffusion through solids involving two parallel paths. We conclude by considering 'pseudo' steady state problems, in which the boundary conditions are not fixed, but change only slowly with time.

Diffusion tends to move a system towards one of two stationary states, denoted *equilibrium* and *steady state*, depending on the imposed boundary conditions. If, for example, we add a few drops of a dye into a clear liquid, diffusion occurs until the dye is uniformly distributed throughout the liquid. This represents an equilibrium state. In concrete terms it is represented by a uniform concentration (or by zero concentration gradient) everywhere in the body. If, however, we separate a chamber filled with hydrogen at high pressure from another which is empty of hydrogen by a thin membrane (say of steel), then after some time we will establish a steady state in which the amount of

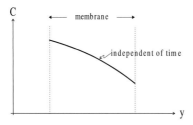

Figure 2.1 A schematic illustration of steady-state diffusion.

hydrogen diffusing through the membrane per unit time is a constant. Within the membrane, the concentration gradient is not zero. However, once a steady state is established, *the concentration at any location does not change with time.* This defines a steady state, as illustrated in Fig. 2.1.

Fick's First Law is most useful in dealing with problems involving a steady state, once steady-state conditions have been reached, and we will consider several examples in this chapter.

2.1   Fick's First Law

We were introduced to Fick's First Law in the previous chapter as an empirical equation of the form

$$J_B = -D_B \frac{\partial C_B}{\partial y} \tag{2.1}$$

Here J_B is the flux of solute B while C_B is the concentration of this solute. The equation is clearly written for a concentration gradient along the y-axis only.

It is a straightforward matter to extend Fick's First Law to three dimensions by replacing $\partial C_B/\partial y$ by the full gradient C_B, *i.e.*

$$\mathbf{J_B} = -D_B \nabla C_B = -D_B \left(\frac{\partial C_B}{\partial x}, \frac{\partial C_B}{\partial y}, \frac{\partial C_B}{\partial z} \right) \tag{2.2}$$

The flux $\mathbf{J_B}$ is therefore a vector quantity.

2.2   Applications to steady-state problems

2.2.1   Measurement of the diffusion coefficient

Let us consider what happens if a slab of some solid is used to separate two chambers. The slab has a thickness d and a cross-sectional area A, as illustrated in Fig. 2.2. We now suppose that by some means we are able to fix the concentration of some solute B at either side of this slab. We will designate the concentration of B as C_1 on one side (at $y = 0$) and as C_2 on the other side (at

$y = d$). If $C_1 > C_2$, then solute will diffuse from left to right and eventually a steady state will be established. We can then measure by some means the amount of solute q which is transported through the slab in a time interval t. The flux is a constant once steady state is established and is equal to $J_B = q/(At)$. We can now slice up the slab and measure the concentration as a function of position from one side of the slab to the other. A hypothetical plot of this is shown in Fig. 2.3. The diffusion coefficient can be obtained from this plot by inverting Fick's First Law, *i.e.*

$$D_B = -\frac{J_B}{\frac{\partial C_B}{\partial y}} = -\frac{q}{At \frac{\partial C_B}{\partial y}}. \tag{2.3}$$

In Fig. 2.3, the slope of the C_B vs y curve changes as a function of y. This is the most general case. A straight line relationship in Fig. 2.3 (for the planar geometry we are currently considering) would indicate that D_B is a constant, independent of position, and therefore of concentration. In this case

$$D_B = \frac{qd}{At(C_1 - C_2)}. \tag{2.4}$$

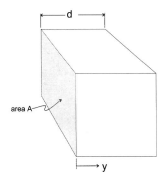

Figure 2.2 We will consider diffusion through a solid slab with cross-sectional area A and thickness d.

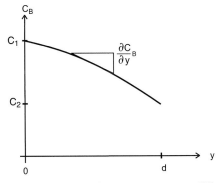

Figure 2.3 At steady state, a constant diffusion profile is developed. The slope $\partial C/\partial y$ at any given point is related directly to the diffusion coefficient.

A classical experiment of this type is that by Smith[¶] who measured the diffusion coefficient of carbon in steel. In this case, however, he used a hollow cylinder instead of a plate, which provides a more practical experimental geometry (see Fig. 2.4). He passed a carburizing gas through the inside of the cylinder, which resulted in a high carbon concentration, say C_0, on the inside wall. Meanwhile, a decarburizing gas was passed over the outside of the cylinder resulting in a low carbon concentration on the outside wall. At steady state the concentration remained constant at each radial position within the cylinder, $a < r < b$. This means that the total amount of carbon passing through any cylindrical element per unit time was the same, *i.e.* $J \cdot 2\pi r \ell$ is a constant; we will call it $\dot{q}$. Here ℓ is the length of the cylinder. The constant $\dot{q}$ is also equal to the amount of carbon added to the decarburizing gas per unit time. This can be measured experimentally (for example, by using a mass spectrometer). Thus $\dot{q}$ is a known experimental constant. We can then substitute Fick's First Law (which in radial coordinates is simply $J = -D\, \partial C/\partial r$) into this equation, and integrate.[†] The result is

$$C - C_0 = \frac{\dot{q}}{2\pi \ell D} \ln \left(\frac{a}{r} \right). \tag{2.5}$$

The constants of integration C_0 and a can be determined from the boundary

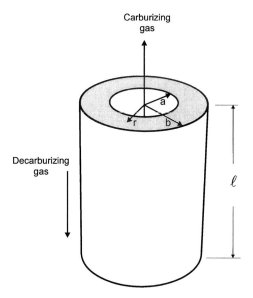

Figure 2.4 An illustration of the classic experiment by R. P. Smith (1953) in which the diffusion coefficient of C in Fe is measured using a cylindrical geometry.

[¶] R. P. Smith, *Acta Metall* **1**, 578, (1953)
[†] Note that, for the sake of clarity, we have dropped the subscript B, in the flux and in the concentration. It is important to realize, however, that this equation applies to a specific diffusing species, in this case carbon.

conditions which set the concentrations at $r = a$ and $r = b$. However if we merely want to determine the diffusion coefficient, we note that this can be obtained by plotting C vs $\ln r$, the slope of which is given by $-\dot{q}/(2\pi\ell D)$. Once again a constant slope would indicate a concentration-independent diffusion coefficient. This is the method used originally by Smith to measure the diffusion coefficient of carbon in iron, D_C in Fe.

Example 2.1: Diffusion of hydrogen through a hollow cylinder

Suppose that a hollow cylinder of nickel is used to measure the diffusion coefficient of hydrogen through this material. The cylinder has an outer diameter of 2 mm and a base (inner diameter) of 1 mm. A CH_4/H_2 mixture is passed through the base of the cylinder which fixes the hydrogen concentration on the inner wall as 0.01wt%. Hydrogen diffusing through the cylinder is evacuated so that the hydrogen concentration on the outer wall is effectively zero. If hydrogen is collected (under steady-state conditions) at a rate of 1.1×10^{-11} moles/s from a cylinder, 2 cm long, held at $200\,°C$, what is the diffusion coefficient?

We first invert eq. (2.5) in terms of the diffusion coefficient:

$$D = \frac{\dot{q}}{2\pi\ell(C - C_0)} \ln\left(\frac{a}{r}\right).$$

If we let $r = b$ the outer radius, then $C = 0$.

We now encounter a problem that occurs frequently in diffusion problems having to do with the conversion of units. The hydrogen concentration at the surface is given as a weight fraction (expressed as wt%). However, the concentration required in Fick's Laws must be expressed in units of matter (either mass or weight) per unit volume. Thus we must convert this concentration. Table 2.1 provides a number of conversions that are useful for this. From the table we see that, in mass units, $C = X^*\rho$ where X^* is the weight fraction of hydrogen in the nickel and ρ is the density of nickel (*i.e.* $\rho = 8790 \text{ kg/m}^3$). From this we find that

$$C_0 = 0.879 \text{ kg/m}^3 (\textit{i.e. kg of } H_2 \text{ per m}^3 \text{ of the solid Ni phase}).$$

We must also convert the flow rate into compatible units of kg/s. To do this we multiply by the molar weight of hydrogen (2×10^{-3} kg/mol). Thus

$$\dot{q} = 2.2 \times 10^{-14} \text{ kg/s}.$$

The other terms are all known. Therefore, by substitution

$$D = 1.38 \times 10^{-13} \text{ m}^2/\text{s}$$

at $200\,°C$.

2.2.2 Permeability of gases through a solid

A common problem in materials engineering pertains to the permeability of gases through solid membranes or films. Quite often these problems involve interstitial elements which diffuse rapidly through many solids. Sometimes, the rapid diffusion of solutes through thin films represents a problem which must be prevented. For example, the diffusion of hydrogen through steel is so rapid that even relatively thick plates can be penetrated in modest times at low temperatures. On other occasions, high gas permeability presents opportunities. For example, the rapid diffusion of oxygen through a zirconia ceramic enables this material to be used as an oxygen sensor at high temperature. When thin films or membranes are involved it is generally not practical to section them and measure the concentration profile directly. Moreover, we are really interested in knowing how much gas will flow through a membrane per unit time. This turns out to be a function of both the diffusivity and the

Table 2.1 Converting between different units of concentration.

Quite often we need to convert compositional data for use in diffusion equations. For example, phase diagrams often give us compositions as weight fraction, or sometimes mole fraction. The diffusion equations, however, demand concentrations in units of moles or weight per unit volume. Given below are a series of conversions that are of value in this regard:

Concentration (mass/unit volume), $C_M = X^* \cdot \rho$
Concentration (moles/unit volume), $C_N = X/\Omega = XC$,

where

X^* is the weight fraction of the species
X is the mole fraction of the species
Ω is the molar volume of the phase
C is the overall concentration of a phase (mol/vol); $C = 1/\Omega$.

Similarly, we often need to convert the amount of a given phase present between weight and volume fraction units. For this:

Volume fraction, $\varphi = f^* \rho_a / \rho$,

where

f^* is the weight fraction of the phase
ρ is the density of the phase
ρ_a is the average density of the alloy.

The average density of an alloy can be determined from that of the individual phases using either

$$\rho_a = \sum_i \rho_i \varphi_i \quad \text{or} \quad 1/\rho_a = \sum_i f^*_i / \rho_i$$

Note: the molar volume is equal to the molar weight divided by the density, both quantities that are more readily found in tables.

solubility. We can therefore define a single parameter, called the 'perme-ability', which characterizes this phenomenon.

Let us consider the flow of a diatomic gas species (*e.g.* H_2, O_2 or N_2) through a solid membrane of thickness δ, as illustrated in Fig. 2.5. The gas pressure on the left side of the membrane is fixed at p_1. Assuming that chemical equilibrium is established locally between the gas and the solid surface, the surface concentration of the dissolved gas will have a well-defined value C_1. Similarly, the gas pressure on the right side is fixed at p_2 resulting in a surface concentration of the dissolved gas equal to C_2. Once steady state is established we can measure the number of moles q of the gas which enters the right chamber during a time t. If the cross-sectional area of the membrane is A then the flux is equal to

$$J = \frac{2q}{At}. \tag{2.6}$$

The factor of two arises since diatomic gases usually dissolve into solids in atomic form. Thus, there are two moles of diffusing atoms for every mole of gas coming from the right surface of the membrane. Based on Fick's First Law we can also write for the flux

$$J = -D\frac{\partial C}{\partial y} = +D\frac{C_1 - C_2}{\delta} \tag{2.7}$$

In order to proceed we need to relate the surface concentration of dissolved gas to the gas pressure. This can be determined, as noted above, by assuming that the surface is in *local equilibrium* with the gas. This is an assumption that we will often use and one which is generally valid. We will examine this in more detail later, starting in Chapter 4. For a diatomic gas B_2 the equilibrium reaction at the surface is

$$B_{2(g)} \rightleftharpoons 2\underline{B}, \quad \Delta G^{\circ} = -RT \ln K. \tag{2.8}$$

The underline is used to symbolize a dissolved species. If we denote the equilibrium constant for this reaction by K then the equilibrium concentration of $\underline{B}$

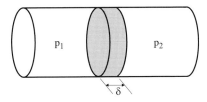

Figure 2.5 A diatomic gas flow from a chamber at high pressure p_1 (left), through a membrane of thickness δ, to a chamber at a lower pressure p_2 (right).

dissolved in the solid is related to the pressure of B_2 in the gas by

$$X_B = \sqrt{K \cdot p_{B_2}} \qquad (2.9)$$

where X_B is the molar fraction of B. This is related to the concentration C_B (in moles/unit volume) by the relation $C_B = X_B/\Omega$ (see Table 2.1), where Ω is the molar volume. We can therefore write the surface concentrations as $C_1 = K^*\sqrt{p_1}$ and $C_2 = K^*\sqrt{p_2}$ respectively, where K^* is the solubility (in moles per unit volume) at 1 atm pressure (*i.e.* $K^* = \sqrt{K/\Omega}$). Therefore, the flux through the solid is equal to

$$J_B = \frac{D_B}{\delta} K^*(\sqrt{p_1} - \sqrt{p_2}). \qquad (2.10)$$

From this equation we see clearly that the permeability of a gas through a solid depends on the product of the diffusivity D_B and the equilibrium solubility of the gas in the solid, K^*. This leads to one definition of permeability[¶] as this product, *i.e.* $\Pi = D_B \cdot K^*$. The gas permeability can be determined by measuring q, the amount of gas penetrating the membrane during time t, once steady-state conditions have been established. If we equate the two flux equations shown above (eqs. (2.6) and (2.10)) then permeability is found from q using

$$\Pi = \frac{2\delta q}{(\sqrt{p_1} - \sqrt{p_2})At}. \qquad (2.11)$$

According to this relationship, permeability should have units of moles/ $\text{m s atm}^{1/2}$. However, these are not the units conventionally used in this field. Instead of defining solubility as moles/m^3 in the solid for 1 atm gas pressure, it is defined as the volume of gas (at standard temperature and pressure – STP) which is dissolved per unit volume of solid; *e.g.* m^3 (STP)/m^3. This makes the units of flux equal to m^3 (STP)/m^2 s, and that of permeability equal to m^3 (STP)/m s atm$^{1/2}$.

The permeability depends on two thermally activated processes (diffusivity and solubility) and can therefore be written in the form $\Pi = \Pi_0 \exp(-Q_p/RT)$. Some typical data are shown in Table 2.2. We can use this data to show why steel pressure vessels are often coated with copper to decrease their permeability to hydrogen. At room temperature the permeability of hydrogen through copper is about 10^8 times less than that of hydrogen through steel. This arises primarily due to the low solubility of hydrogen in copper. Thus even a very thin film of copper on the surface of a thick steel pressure vessel greatly improves the hydrogen retention (provided that the copper layer integrity is not compromised by scratches or pitting).

[¶] Unfortunately, there are several different definitions of permeability in current use so one needs to be careful when using any permeability data that is found in books or articles.

Finally, before we leave this topic, let us examine the assumption of local equilibrium at the gas/metal interface. We can test whether this is actually valid experimentally, by measuring the flux of gas passing through the solid as a function of the membrane thickness δ. According to eq. (2.10), the flux should be inversely proportional to δ. We might expect this assumption to break down as δ gets very small, since the flux may increase to the point where surface reaction rates are slower than the diffusion flux and thus dominate the kinetics.

Example 2.2: Solubility and diffusion contributions to permeability

(a) *What is the ratio in permeability for hydrogen in Cu and Fe at room temperature?*

(b) *How much of this ratio is due to diffusion effects and how much to solubility?*

(a) From Table 2.2,

$$\Pi_{H_2-Cu}/\Pi_{H_2-Fe} = \frac{3.35 \times 10^{-14}}{5.58 \times 10^{-6}} = 6.0 \times 10^{-9}.$$

(b) From Appendix B, we can find the diffusivities:

$$D_{H_2-Cu} = 1.96 \times 10^{-7} \exp\left(-\frac{28,800}{8.314 \times 298}\right) = 1.75 \times 10^{-12}\,m^2/s,$$

$$D_{H_2-Fe} = 2.30 \times 10^{-9} \exp\left(\frac{6,600}{8.314 \times 298}\right) = 1.6 \times 10^{-10}\,m^2/s.$$

Thus the ratio of diffusivities is 1.09×10^{-2}. Since $\pi = DK^*$, the ratio of solubilities must be $6.09 \times 10^{-9}/1.09 \times 10^{-2} = 5.48 \times 10^{-7}$. Thus, most of the decreased permeability to hydrogen exhibited by copper is due to a much lower solubility.

Table 2.2 Permeability data for gas–metal reactions.

Gas	Metal	Π_0 m^3 (STP)/m s atm$^{1/2}$)	Q_p (kJ/mol)	Π at 293 K
H_2	Ni	12	57.8	5.95×10^{-10}
H_2	Cu	1.5–2.3	67–78	3.35×10^{-14}
H_2	α-Fe	29	35.1	5.58×10^{-6}
H_2	Al	$3.3–4.2 \times 10^3$	129	3.82×10^{-20}
N_2	Fe	45	99	1.01×10^{-16}
O_2	Ag	29	94	5.06×10^{-16}

Source: D. H. Poirier and G. G. Geiger, *Transport Phenomena in Materials Processing* (TMS, Warrendale, PA, 1994), p. 466.

Example 2.3: Effective permeability of a coated pressure vessel

Suppose that a steel pressure vessel destined for hydrogen storage is to be coated with a thin layer of copper in order to reduce the loss of hydrogen. The uncoated steel vessel has a 15 mm wall thickness, and the hydrogen pressure in the vessel is 2 MPa.

(a) Suppose the copper is deposited on the inside wall of the pressure vessel. What thickness of copper is required in order to reduce the hydrogen loss by 10^6 times as compared with that for an uncoated vessel?
   Hint: You can assume that in the coated vessel all of the pressure loss is across the copper layer.

(b) If you were asked to design a copper-coated pressure vessel, would you recommend that the copper coating be applied to the inside or outside surface of the vessel? State the reasons for your choice.

(a) We will assume that the diameter of the pressure vessel is considerably larger than the wall thickness. Therefore the problem involves a planar geometry, with the diffusive flux normal to the wall. We will also assume that steady-state conditions are established quickly. We can define the gas pressure inside the vessel to be $p^*(= 2\,\mathrm{MPa}$ in this case), and outside to be p°. In doing the evaluation, we will assume that $p^\circ \approx 0$.

From Fick's First Law, as modified in eq. (2.10)

$$J = \frac{D}{L} K^* (\sqrt{p^*} - \sqrt{p^\circ}).$$

Here, D and K^* refer to steel for an uncoated vessel, and to copper for a coated vessel. Similarly, $L = d$, the wall thickness for an uncoated vessel, while $L = \delta$, the thickness of the copper layer for a coated vessel. For the uncoated vessel,

$$J_u = \frac{D_{Fe} K^*_{Fe}}{d} \sqrt{p^*} = \frac{\Pi_{Fe}}{d} \sqrt{p^*},$$

while for the coated vessel,

$$J_c = \frac{D_{Cu} K^*_{Cu}}{\delta} \sqrt{p^*} = \frac{\Pi_{Cu}}{\delta} \sqrt{p^*}.$$

Since we require that $J_c = 10^{-6} J_u$,

$$\frac{\delta}{d} = 10^6 \frac{\Pi_{Cu}}{\Pi_{Fe}}.$$

From Table 2.2, evaluating at 293 K,

$$\Pi_{Cu} = 3.35 \times 10^{-14}\,\mathrm{m}^3(\mathrm{STP})/\mathrm{m\,s\,atm}^{1/2}$$

and

$$\Pi_{\text{Fe}} = 5.58 \times 10^{-6} \, \text{m}^3(\text{STP})/\text{m s atm}^{1/2}$$

$$\therefore \delta = 10^{+6}(1.5 \times 10^{-2}) \, \frac{3.35 \times 10^{-14}}{5.58 \times 10^{-6}}$$

$$= 9 \times 10^{-5} \, \text{m}$$

Therefore, only a thin (90 μm) copper layer is required to greatly reduce the hydrogen gas loss from the vessel by six orders of magnitude.

(b) It is better to coat the inner wall for several reasons. First, a thin layer plated onto the outside of the vessel is liable to wear or abrasion which might leave some regions uncoated. Because of the large difference in permeability, even small gaps in the coating will greatly increase losses. Second, an external coating enables the walls of the vessel to fill with hydrogen. This represents a loss of hydrogen available as a gas. Third, some grades of steel can become embrittled by hydrogen.

2.2.3   Diffusion in parallel through a composite solid

We often wish to consider what happens when a solute diffuses through more than one phase simultaneously. For the moment, we will restrict this to thinking about diffusion through two phases in parallel. By this I mean that the solute can diffuse from one boundary to the other by going through either phase. This is illustrated in Fig. 2.6. We assume that the boundary conditions

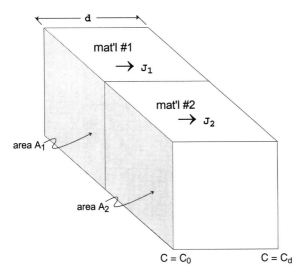

Figure 2.6 A schematic illustration of diffusion through a composite solid. Matter diffuses in parallel through one material with diffusivity D_1, and cross-sectional area A_1, and through another material with diffusivity D_2 and cross-sectional area A_2.

at each side are the same for each phase, *i.e.* $C = C_0$ at $y = 0$, and $C = C_d$ at $y = d$. The cross-sectional areas of the two phases are A_1 and A_2 respectively. Since the diffusion coefficients in each phase are different, *i.e.* D_1 and D_2 respectively, the fluxes will also be different. The total flow rate I, however, can be easily calculated. It is

$$I = J_1 A_1 + J_2 A_2 = -D_1 A_1 \frac{\partial C}{\partial y} - D_2 A_2 \frac{\partial C}{\partial y} \tag{2.12}$$

which, for constant diffusion coefficients at steady state, can be rewritten as

$$I = -(D_1 A_1 + D_2 A_2) \frac{C_d - C_0}{d}. \tag{2.13}$$

If the total cross-sectional area is $A = A_1 + A_2$, then the average flux is given by

$$J = \frac{I}{A} = -\left(D_1 \frac{A_1}{A} + D_2 \frac{A_2}{A}\right) \frac{C_d - C_0}{d} = -D_{\text{eff}} \frac{\partial C}{\partial y}. \tag{2.14}$$

Thus it appears that the diffusion behaviour of a composite solid in parallel can be represented by Fick's First Law provided that we use an effective diffusion coefficient which is simply the average of the diffusion coefficient of each phase weighted by the fraction of the total cross-sectional area which it occupies, *i.e.*

$$D_{\text{eff}} = D_1 \frac{A_1}{A} + D_2 \frac{A_2}{A}. \tag{2.15}$$

The assumption which we made at the outset, to the effect that the boundary conditions for each phase were equal, may seem somewhat artificial and in many cases it is. However, this model can be applied to several important situations in which so-called short-circuit diffusion paths are available. Perhaps the most important of these involves grain boundary diffusion. Grain boundaries represent regions of disorder in a crystal. They therefore lead to rapid diffusion. However, the width of a grain boundary is very small. It is of atomic dimensions, about 0.2 nm. In contrast, diffusion through the crystalline lattice is much slower. However, almost all of a polycrystal consists of crystalline regions. We are therefore left to ask how important is grain boundary diffusion and under what conditions might it control the overall transport properties of a polycrystalline solid? This situation is illustrated in Fig. 2.7(a).

In order to proceed we need to develop a simple picture of a polycrystal from the point of view of diffusion. This is what is known as a 'conceptual model'. (See Appendix D for a discussion on the development of such a model.) In this case the important elements of a polycrystal are that it contains interconnected regions of grain boundaries through which matter can diffuse without entering the lattice. Atoms can therefore diffuse entirely by a grain boundary path or by a lattice path. For our purposes it is sufficient to think of cube-shaped grains of width w, separated by thin grain boundaries of width δ_b, as shown in Fig. 2.7(b). In general $w \gg \delta_b$, *i.e.* δ_b is less than 1 nm while w is generally larger than 1 µm. In a polycrystal with a total cross-sectional area A and containing N grains over this area (see Fig. 2.7(b)) the cross-sectional area of the lattice regions is $A_\ell = Nw^2$. The cross-sectional area of the grain boundaries is $A_b = N \cdot 4w\delta_b/2 = 2N\delta_b w$. We arrive at this by adding up the area around the four sides of each grain, noting that every grain boundary is shared by two grains. We can now find the effective diffusion coefficient from the formula calculated above. It is

$$D_{\text{eff}} = \frac{A_\ell}{A} D_\ell + \frac{A_b}{A} D_b = D_\ell + 2\frac{\delta_b}{w} D_b. \tag{2.16}$$

In deriving this we were able to simplify the result by using $w \gg \delta_b$. From this result we can see that D_{eff} will be dominated by lattice diffusion only if D_ℓ is large with respect to $2(\delta_b/w)D_b$. Since D_b is so much larger than D_ℓ, this will not always be true. The data for grain boundary diffusivity shows that because of the ease with which diffusion occurs it has a lower activation

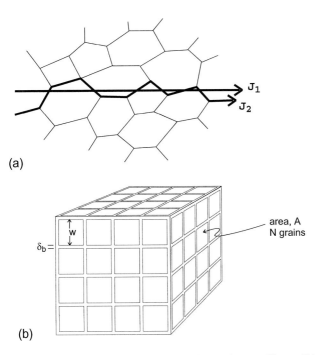

(a)

(b)

Figure 2.7 (a) Atoms diffusing through a polycrystalline solid can use either a lattice or a grain boundary path. (b) A conceptual model of a polycrystalline solid used to develop an equation for the effective diffusion coefficient in a polycrystal due to both lattice and grain boundary diffusion.

energy than lattice diffusion, *i.e.* in the Arrhenius equation for diffusion $D = D_0 \exp(-Q/RT)$, Q is lower for boundary diffusion than for lattice diffusion. Typically Q_b/Q_ℓ is between 0.5 and 0.75. This means that the difference between lattice and boundary diffusivities is greater at lower temperatures. This is shown schematically in Fig. 2.8.

We can determine the temperature at which boundary and lattice diffusion make an equal contribution to the effective diffusivity of a polycrystal by solving for temperature in $w \cdot D_\ell = 2\delta_b \cdot D_b$. If we substitute the Arrhenius equations for lattice and boundary diffusion into this expression we find that

$$\frac{wD_{0\ell}}{2\delta_b D_{0b}} = \exp\left(\frac{Q_\ell - Q_b}{RT}\right). \tag{2.17}$$

We can now solve this for temperature:

$$T = \frac{Q_\ell - Q_b}{R \ln\left(\dfrac{wD_{0\ell}}{2\delta_b D_{0b}}\right)}. \tag{2.18}$$

Consider the following data for pure nickel:

- $D_\ell = D_{0\ell} \exp(-Q_\ell/RT)$,
  with $D_{0\ell} = 1.9 \times 10^{-2}\,\mathrm{m^2/s}$, and $Q_\ell = 284\,\mathrm{kJ/mol}$
- $\delta_b D_b = \delta_b D_{0b} \exp(-Q_b/RT)$,
  with $\delta_b D_{0b} = 3.5 \times 10^{-15}\,\mathrm{m^3/s}$, and $Q_b = 115\,\mathrm{kJ/mol}$

A typical grain size w for a pure metal polycrystal is about 50 μm. Substituting into eq. (2.18) gives a temperature of 1086 K. Thus, for this grain size, grain boundary diffusion is dominant at a temperature below about 1086 K while lattice diffusion is dominant above this temperature. Note that the melting temperature of nickel is 1726 K. If the grain size is increased then this critical

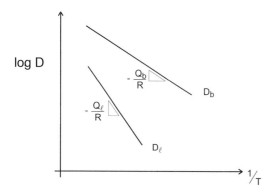

Figure 2.8 An Arrhenius plot (*i.e.* log(rate) vs $1/T$) for diffusion, illustrating that grain boundary diffusion is faster but with a lower activation energy Q, than lattice diffusion.

temperature is decreased and lattice diffusion becomes more important. If the grain size is decreased, the opposite occurs.

Now consider the situation for Al_2O_3. Mass transport is dominated by oxygen ion diffusion, for which:

- $D_{0\ell} = 0.19\,m^2/s$, and $Q_\ell = 636\,kJ/mol$
- $\delta_b D_{0b} = 1.0 \times 10^{-8}\,m^3/s$, and $Q_b = 380\,kJ/mol$

For a typical grain size of $5\,\mu m$ in a ceramic material, the critical temperature we calculate is 7970 K, which is far above the melting temperature. This means that for all temperatures in solid alumina and all reasonable values of the grain size, grain boundary diffusion is dominant. This is typical of many ceramic materials.

There are other instances of short-circuit diffusion besides the one we have just discussed. For example, the cores of edge dislocations can act as rapid paths for diffusion. This can be important during the recovery of dislocations following cold work or during creep. The method we use to analyze this problem is similar to that for the case we have just discussed and is left as an exercise.

2.3   Growth of surface layers – pseudo steady state

There are numerous examples in materials engineering in which films are grown on surfaces by diffusion or similar processes. We will consider two cases – oxidation and slip casting.

2.3.1   Growth of oxide films

When a metal surface is exposed to an oxidizing atmosphere (*e.g.* air) at a sufficient temperature, an oxide film will nucleate and grow on the surface. The details of this process can be quite complex, especially in systems such as iron which exhibit several different metal oxide phases. However, the basic growth process is the same. We will illustrate this by considering nickel which has a single binary oxide, NiO. When exposed to oxygen a NiO oxide film will develop on the surface of the metal, as illustrated in Fig. 2.9. The continued growth of this film requires diffusion through the NiO. There are two species that can diffuse here. Ni^{2+} ions can diffuse from the metal/oxide interface to the outer surface where they will react with the oxygen in the atmosphere to create more NiO. Alternatively, O^{2-} ions can diffuse from the surface to the metal/oxide interface where they will react with the nickel to form more NiO.[¶]

[¶] Strictly we should consider the diffusion of a third species, namely electrons. Since both nickel and oxygen diffuse in NiO as ions a counter flow of electrons is required to maintain charge neutrality. Generally this occurs rapidly enough that it does not impede the growth of the film.

These two processes for adding to the thickness of the film act independently of one another. Therefore, which ever is the faster will control the overall rate of film growth. It this case nickel has the higher diffusion coefficient and the NiO film will grow primarily at the outer surface. In order to understand how this film grows by diffusion we first need to realize that NiO, like all compounds, exists with a range of *stoichiometry*. Thus the Ni/O ratio in NiO is slightly higher when the film is in contact with nickel than when it is in contact with oxygen. There will therefore be a gradient in the nickel concentration within the film, as illustrated in Fig. 2.10(a). We assume that the film is locally in equilibrium with pure nickel on one side and oxygen at a given partial pressure p_{O_2} on the other side. The nickel concentration in each case is given by C_1 and C_2 respectively. We will also assume that the diffusion coefficient is independent of concentration within the NiO.

Let the current thickness of the film be equal to L. As time goes on and diffusion proceeds, this thickness will increase. We can therefore never reach a true condition of steady state since we defined this by saying that the concentration at any position is not varying with time. As Fig. 2.10(b) shows this is never true. However, if the film grows slowly enough it will behave *as if it were at steady state*. What this means is that when the film grows slowly with respect to the rate of diffusion the concentration profile is able to keep up with the steady-state profile. We call this condition a 'pseudo' steady state. In chapter 6, where we analyze transient problems, we will return to this situation and show exactly what conditions must be met in order for pseudo-steady-state conditions to apply. However, we find that this assumption is always valid for the growth of films when they are very thin.

With this assumption we can write down the flux of Ni^{2+} ions as

$$J = -D \frac{\partial C}{\partial y} = -D \frac{C_2 - C_1}{L(t)}.$$ (2.19)

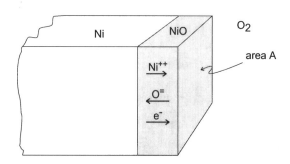

Figure 2.9 The growth of an oxide scale on a metal surface (*e.g.* NiO on Ni) involves the diffusion of metal ions to the outer surface and oxygen ions to the metal/oxide interface. The faster species will control the overall growth rate of the scale.

As the nickel ions arrive at the surface they react with oxygen and produce NiO. We can use a mass balance to determine the relationship between the flux and the rate of thickening. If the flux is in units of moles/unit area/unit time, then

$$J = \frac{1}{\Omega_{\mathrm{NiO}}} \frac{\mathrm{d}L}{\mathrm{d}t}, \tag{2.20}$$

where Ω_{NiO} is the molar volume of NiO.[1] If we equate these two expressions for the flux, then

$$L\,\mathrm{d}L = -D\Omega_{\mathrm{NiO}}(C_2 - C_1)\,\mathrm{d}t. \tag{2.21}$$

We can integrate this from zero time (when $L = 0$) up to the current time and

(a)
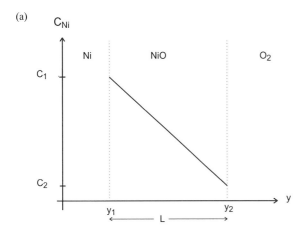

(b)
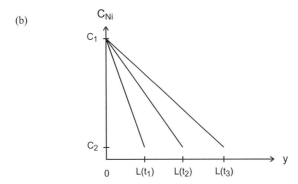

Figure 2.10 (a) Diffusion across an oxide scale is driven by the difference in the metal/ oxygen ratio between the metal/oxide interface and the outer surface. (b) As the film grows, C_1 and C_2 remain fixed by local equilibrium conditions. If the film grows slowly, the concentration profile is close to that for steady-state conditions. We call this condition, 'pseudo' steady state.

[1] The molar volume is equal to the molecular weight (for NiO this is 74.7g/mole) divided by the density (for NiO this is 6.72g/cm^3). Then the molar volume of NiO is 11.1 cm^3/mol or 1.11×10^{-5} m^3/mol.

solve for the thickness of the film. The result is

$$L(t) = \sqrt{2D\Omega_{\text{NiO}}(C_1 - C_2)t}. \tag{2.22}$$

This is the classical equation for the parabolic growth of a film, since $L \propto \sqrt{t}$. If an oxide film continues to grow by this process its rate of thickening decreases steadily as the diffusion distance increases. It eventually becomes so slow that it effectively stops. We call this kind of film growth 'passivating'. The most common example occurs on aluminum. For many materials, however, the oxide films break or spall as they thicken. The rate of oxidation then increases substantially and is strongly influenced by other factors such as the local microstructure underneath the breaks in the film.

Example 2.4: Oxidation of nickel alloys

Nickel alloys are used in gas turbine engines at temperatures in excess of 1000 °C. If pure nickel is exposed to air at 1000 °C it will develop a coherent oxide scale. Determine how thick the oxide scale will be after 1, 10 and 100 hours.

In order to answer this question we need to know how the stoichiometry of NiO changes with oxygen potential. From this we can determine the boundary conditions. According to Kofstad,[¶] NiO becomes metal deficient, $Ni_{1-x}O$, as the oxygen-potential is increased. This takes the form of Ni^{2+} vacancies. In equilibrium with air at 1000 °C, $x = 3 \times 10^{-3}$. In equilibrium with Ni metal, x is much lower, effectively zero (*i.e.* at the Ni/NiO interface NiO is approximately stoichiometric). Therefore the Ni^{2+} vacancy concentration difference across the NiO scale is $X_1 - X_2 = 3 \times 10^{-3}$. The Ni^{2+} ion diffuses more quickly in NiO than the O^{2-} ion. Therefore, Ni^{2+} diffusion will dominate oxidation.[†] At 1000 °C, $D = 2.5 \times 10^{-15}\,\text{m}^2/\text{s}$. To complete the problem we need to convert from X, the mole fraction of Ni ions (or Ni vacancies), to concentration C, using (see Table 2.1), $X = C\Omega_{\text{NiO}}$. Therefore, from eq. (2.22), the scale thickness is given by

$$L(t) = \sqrt{2D(X_1 - X_2)t}.$$

The result is:

Time (h)	Scale thickness L (nm)
1	232
10	735
100	2320

[¶] P. Kofstad, *Nonstoichiometry, Diffusion and Electrical Conductivity in Binary Metal Oxides* (McGraw-Hill, New York, 1972)
[†] This is also important to the stability of the oxide scale. Since NiO occupies about 60% more volume than Ni, if the scale forms at the Ni/NiO interface, the volume expansion would cause stresses and lead to spalling. However, since the film forms at the outer surface, the volume expansion is more easily accommodated. In contrast, the oxidation of Ti is controlled by O^{2-} diffusion. Thus, despite the smaller expansion (50%), TiO_2 scales are powdered and spall off. This limits the use of Ti alloys at elevated temperatures.

2.3.2 Slip casting

A similar problem to the one we have just discussed occurs in slip casting of ceramics. This is an important process for making a wide variety of ceramic materials, from sanitary ware (*e.g.* glazed porcelain sinks) to automotive engine components (*e.g.* spark-plug insulators). The process involves suspending the ceramic particles in water along with other chemicals such as dispersants and binding agents. This mixture is called a 'slip.' The slip is then poured into a gypsum mould. Gypsum is a porous material which contains a network of fine interconnected pores (see Fig. 2.11(a)). When the slip is poured into the gypsum mould the water is drawn into the pores by capillary forces. But the

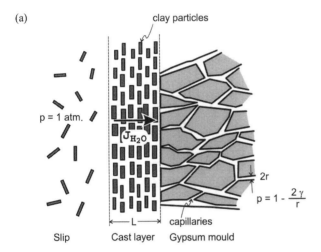

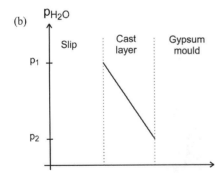

Figure 2.11 (a) An illustration of the slip casting process for forming ceramic bodies. Water is drawn out of the slurry by the capillary pressure in the gypsum mould (after Kingery and Bowen, 1976).[¶] (b) Water diffuses through the cast layer due to a pressure gradient.

[¶] Readers wishing more details on this process and its analysis are referred to W. D. Kingery and H. K. Bowen, *Introduction to Ceramics* (Wiley and Sons, 1976), pp. 385–6

ceramic particles are too large. They are therefore deposited (*i.e.* cast) onto the surface of the mould. The thickness of the cast layer increases with time. Once it is thick enough, the remaining slip is poured out of the mould. The cast material is then dried and fired to produce a finished ceramic product. At first this may seem to have little relationship to the diffusion problems we have been discussing. However, the rate of slip casting is determined by the rate at which water is able to diffuse through the cast layer. This is governed by a modified version of Fick's First Law

$$J = -k_s \frac{\mathrm{d}p}{\mathrm{d}y} \tag{2.23}$$

in which the concentration is replaced by pressure p and the rate constant k_s is related to, but not exactly equal to, the diffusion coefficient. A pressure gradient exists because the pressure in the slip is equal to 1 atm., while the capillary pressure in the gypsum pore channels p_2 is $(1 - 2\gamma/r)$ atm (see Fig. 2.11(b)). Here γ is the interfacial energy per unit area between gypsum and water, while r is the radius of the pore channels. Since γ is of order 1 J/m^2 and the pore-channel radius can be as fine as 1 μm, this produces large negative pressures of order 2×10^6 Pa (*i.e.* 20 atm.). The large pressure gradient drives water through the deposited layer. As this occurs, ceramic particles are deposited on the outer surface of the layer at a rate governed by the flux of water. Thus $\mathrm{d}L/\mathrm{d}t \sim J$, as before. This leads once again to a pseudo-steady-state situation, in which a parabolic relationship of the form

$$L = \sqrt{k_s' t} \tag{2.24}$$

is found. Here k_s' is a rate constant that will depend on k_s and other parameters in a similar manner to the oxidation case.

2.4 Further reading

P. Shewmon, *Diffusion in Solids*, 2nd edition (TMS-AIME, Warrendale, PA, 1989)

J. E. Shelby, *Handbook of Gas Diffusion in Solids and Melts* (ASM International, Materials Park, OH, 1996)

2.5 Problems to chapter 2

2.1 Define what is meant by the term, 'pseudo steady state'. Give at least two examples of problems in which this concept can be used to develop models of diffusion behaviour.

2.2 The self-diffusion coefficient for pure elements may be written as

$$D_\ell = D_{0\ell} \exp(-Q_\ell/RT),$$

where the subscript ℓ represents diffusion through the lattice. Note that T is the absolute temperature and R is the gas constant. In addition, it is possible for atoms to diffuse by one of several short-circuit paths such as the grain boundary and dislocation cores. We derived an expression in Section 2.2.3 which accounts for the effect of grain boundary diffusion. Derive a similar expression for the effective diffusion coefficient in terms of combined lattice and dislocation-core diffusion. In setting up your model, assume that the material contains a known density of dislocations, ρ (in numbers of dislocations per unit area). In addition, you may assume that each dislocation core provides an effective diffusion path of area b^2, where b is the magnitude of Burger's vector of the dislocation. The diffusion coefficient for core diffusion will have a similar form to that for lattice diffusion given above.

For pure nickel:

$$b = 2.5 \times 10^{-10}\,\text{m}$$

$$D_{0\ell} = 1.9 \times 10^{-4}\,\text{m}^2/\text{s}$$

$$D_{0c} = 5 \times 10^{-4}\,\text{m}^2/\text{s}$$

$$Q_\ell = 284\,\text{kJ/mol}$$

$$Q_c = 170\,\text{kJ/mol}$$

Estimate the temperature at which dislocation-core diffusion and lattice diffusion contribute equally for a dislocation density of (a) $10^{10}/\text{m}^2$, and (b) $10^{12}/\text{m}^2$.

2.3   Hydrogen is stored at 1 MPa pressure, at 400 °C, in a spherical steel vessel of 0.5 m radius. The wall thickness of the vessel is 10 mm. Take the diffusion coefficient for hydrogen in iron at this temperature to be $10^{-8}\,\text{m}^2/\text{s}$, and assume that the hydrogen concentration at the vessel walls is in equilibrium with the gas pressure acording to

$$X(p) = 10^{-5}\sqrt{p}$$

where X is the weight fraction of H in the steel, and p is the pressure in MPa.

Develop an expression for the pressure in the vessel as a function of time, once pseudo-steady-state conditions are attained. Determine a value for the rate of pressure drop when the pressure in the vessel is still close to 1 MPa. Why does this problem involve a 'pseudo' steady state?

2.4   During the early growth of an oxide scale on the surface of a piece of cobalt, the scale is polycrystalline, with columnar grains running all the way through the scale (*i.e.* each grain is in contact with both the metal and the gas surface).

(a) Starting with Fick's Law, derive an equation for the thickness of the scale as a function of time, assuming that a 'pseudo' steady state has been established. As part of your anaswer, indicate what the boundary conditions will be (as precisely as you can), and draw schematic curves for the concentration profiles.

(b) Using a simple conceptual model (see appendix D), derive an expression for the effective diffusion coefficient, for diffusion *through* the scale. Will the diffusion coefficient in the scale parallel to the surface be the same as this or different, and why?

Chapter 3

Transient diffusion problems

It was a saying of the ancients that 'truth lies in a well';
and to carry on the metaphor, we may justly add, that
logic supplies us with steps whereby we may go down to
reach the water.

Isaac Watts

Fick's First Law is still valid even when steady-state conditions do not exist. However, it is not very convenient to use it in this form for solving transient diffusion problems, for which the concentration at any position depends on both time and position. It is therefore more convenient to derive a second version of this equation, Fick's Second Law, which contains an explicit dependence on time. In this chapter we will first derive the equation; a second-order partial differential equation. We will then learn how solutions to this equation can be obtained for a large variety of situations depending on the problem geometry, the initial and boundary conditions and the time-frame over which we need a valid solution.

3.1 Fick's Second Law

We consider the situation illustrated in Fig. 3.1. At two positions along the y-axis at y_1 and y_2, a distance Δy apart, the concentration of the solute is C_1 and C_2 respectively. We can use Fick's First Law to determine the amount of solute per unit time which enters this element at y_1 and which leaves it at y_2. Because matter must be conserved, the difference between these must be equal to the rate at which solute accumulates within the element. Writing this as a word equation, we have

$$\frac{\partial C}{\partial t} = \frac{\text{no. of moles entering per unit time}}{\text{volume}}$$
$$- \frac{\text{no. of moles leaving per unit time}}{\text{volume}}. \tag{3.1}$$

If the element has a cross-sectional area A, we can now write this in terms of fluxes as

$$\frac{\partial C}{\partial t} = \frac{AJ_1 - AJ_2}{A \, \Delta y} = \frac{J_1 - J_2}{\Delta y} = -\frac{\partial J}{\partial y}. \tag{3.2}$$

This says that the rate at which the concentration changes with time, at a given location, is equal to the spatial gradient of the flux. We can now substitute Fick's First Law into this equation, which gives

$$\frac{\partial C}{\partial t} = -\frac{\partial}{\partial y}\left(-D\frac{\partial C}{\partial y}\right) = \frac{\partial}{\partial y}\left(D\frac{\partial C}{\partial y}\right). \tag{3.3}$$

We call this equation *Fick's Second Law*. You should note that it contains no new physical information since we derived it from Fick's First Law with the use of conservation of matter only. It is just an alternative form of Fick's First Law which contains an explicit dependence on time.

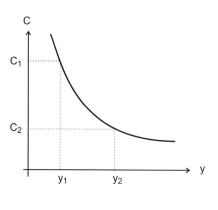

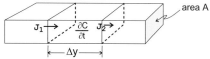

Figure 3.1 A schematic concentration profile, used to illustrate the derivation of Fick's Second Law. If the solute flux entering a small section of length Δy at y_1 is different from that leaving the section at y_2, then the solute concentration must be changing with time.

If the diffusion coefficient does not depend on the concentration of the diffusing species (and therefore on position) Fick's Second Law can be simplified by taking D out of the inner differential. It then becomes

$$\frac{\partial C}{\partial t} = D\frac{\partial^2 C}{\partial y^2}. \tag{3.4}$$

We also need to be able to handle transient diffusion problems in three dimensions. To do this we must generalize the derivation we have just completed. We consider a small three-dimensional element in a solid having dimensions of Δx, Δy and Δz, as shown in Fig. 3.2. The difference in flux entering and leaving in each direction results in a changing concentration within the element. Therefore we can write

$$\frac{\partial C}{\partial t} = \frac{J_{x1} - J_{x2}}{\Delta x} + \frac{J_{y1} - J_{y2}}{\Delta y} + \frac{J_{z1} - J_{z2}}{\Delta z}$$

$$= -\left(\frac{\partial J}{\partial x} + \frac{\partial J}{\partial y} + \frac{\partial J}{\partial z}\right)$$

$$= -\nabla \cdot \mathbf{J}. \tag{3.5}$$

Thus, in generalizing from one dimension we have simply replaced the flux gradient dJ/dy by the divergence of the flux $\nabla \cdot \mathbf{J}$, where the raised **bold** dot represents a dot product of the del operator (∇) and the flux. We can now substitute the three-dimensional version of Fick's First Law (eq. 2.2) into this equation. The result is

$$\frac{\partial C}{\partial t} = \nabla \cdot (D\nabla C). \tag{3.6}$$

As before, if the diffusion coefficient is independent of concentration it can be moved outside the parentheses and the equation simplifies to

$$\frac{\partial C}{\partial t} = D\nabla \cdot (\nabla C) = D\nabla^2 C. \tag{3.7}$$

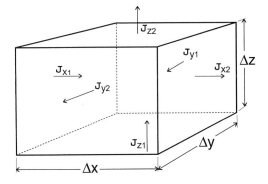

Figure 3.2 An illustration of the fluxes entering and leaving a small three-dimensional element. the build-up of solute is related directly to the divergence of the flux.

The symbol ∇^2 is called Laplace's operator and has the form

$$\nabla^2 C = \frac{\partial^2 C}{\partial x^2} + \frac{\partial^2 C}{\partial y^2} + \frac{\partial^2 C}{\partial z^2}. \tag{3.8}$$

Under steady-state conditions Fick's Second Law (as given by eq. (3.7)) is still valid, of course, but $\partial C/\partial t$ is equal to zero. In this case the equation reduces to

$$\nabla^2 C = 0 \tag{3.9}$$

which is known in mathematics as Laplace's equation. While this equation can be solved, it is usually easier to work with Fick's First Law under steady-state conditions, as we have seen.

3.2   Solutions to Fick's Second Law

3.2.1   Introduction

Fick's Second Law is one of a number of important partial differential equations that occur in engineering. In mathematics textbooks this equation is often referred to as the 'heat equation'. This is because the same equation governs conductive heat transfer. The only difference is in the physical meaning of the variables involved. For conductive heat transfer, the concentration is replaced by the temperature and the mass diffusivity is replaced by the thermal diffusivity (see Table 3.1). Thus, any solution to the heat equation is automatically valid as a solution of Fick's Second Law. Because of its importance in engineering, many solutions to this equation have been developed. A large number of these have been compiled in two classic books on the

Table 3.1 A comparison of governing equations for heat and mass transfer.

Diffusion	Heat conduction
$J = -D\,\partial C/\partial y$ (Fick's 1st Law)	$J = -k\,\partial T/\partial y$ (Fourier's Law)
$D \equiv$ diffusivity	$k \equiv$ thermal conductivity
$\dfrac{\partial C}{\partial t} = \dfrac{\partial}{\partial y}\left(D\,\dfrac{\partial C}{\partial y}\right)$	$\dfrac{\partial T}{\partial t} = \dfrac{1}{\rho c_p}\dfrac{\partial}{\partial y}\left(k\,\dfrac{\partial T}{\partial y}\right)$
$= D\,\dfrac{\partial^2 C}{\partial y^2}$	$= \alpha\,\dfrac{\partial^2 T}{\partial y^2}$

Note: the second version of the transient equations assumes that the material parameter, D or α respectively, is constant, independent of position within the solid. $\alpha \equiv k/(\rho c_p)$, thermal diffusivity; $\rho \equiv$ density; $c_p \equiv$ heat capacity at constant pressure.

subject.[¶] Fick's Second Law falls into the class of partial differential equations generally known in mathematics as 'boundary value problems'. This means that they describe how a system evolves with time within a certain bounded region. In order to solve such problems we need to know more than just the governing equation. We also need to specify:

- initial conditions, which describe the state of the system everywhere within the bounded region at some given initial time, and

- boundary conditions, which describe the state of the system at two specific locations for all time.

These concepts are discussed in further detail in Chapter 4.

3.2.2   Approaches to solving transient diffusion problems

We want to develop a general approach for finding solutions to Fick's Second Law for a wide range of conditions. We first need to determine the geometry of the problem. This is especially important if we want to solve the problem analytically. Why is this? Well a diffusion problem involving an object of any arbitrary shape can *in principle* be solved. For example, we may wish to understand the concentration of nitrogen throughout a steel gearwheel during a nitriding operation (see Fig. 3.3). However, to do this in detail for such a complex geometry will require the use of elaborate numerical methods such as a three-dimensional finite-element program. Many times we do not need an

Figure 3.3 A gearwheel is an example of a geometry which may lead to a complex diffusion problem.

¶ H. S. Carslaw and J. C. Jaeger, *Conduction of Heat in Solids,* 2nd edition (Clarendon Press, Oxford, 1959), and J. Crank, *The Mathematics of Diffusion,* 2nd edition (Clarendon Press, Oxford, 1975)

answer of sufficient detail and accuracy to justify the time and expense involved in such a process. Moreover, the accuracy of the diffusion data that is available is often insufficient to justify such complex solution methods. We therefore rely on analytical solutions that are available for a number of standard geometries. These generally involve one of the simple symmetries: planar, spherical or cylindrical.

This may require that we approximate the real shape of a component by something much simpler. Consider the example of the gearwheel. If the distance the nitrogen diffuses in from the surface (which, we will learn later in this chapter, can be approximated by $\sqrt{Dt}$) is small with respect to the depth of the teeth on the gearwheel then the diffusion flux operates normal to the local surface everywhere and a planar geometry gives an adequate approximation (Fig. 3.4). If however the nitrogen penetrates to the centre of the gearwheel then we can approximate the gearwheel as a cylinder and ignore the teeth (Fig. 3.5).

In each case the spatial dependence in the Laplace operator can be reduced to a single variable. For planar symmetry, as we have already seen, Fick's Second Law becomes

$$\frac{\partial C}{\partial t} = D \frac{\partial^2 C}{\partial y^2}. \tag{3.10}$$

For problems which are spherically symmetrical it becomes

$$\frac{\partial C}{\partial t} = D \left(\frac{\partial^2 C}{\partial r^2} + \frac{2}{r} \frac{\partial C}{\partial r} \right), \tag{3.11}$$

where r is the radial distance from the centre of the object.

For problems exhibiting cylindrical symmetry Fick's Second Law becomes

$$\frac{\partial C}{\partial t} = D \left(\frac{\partial^2 C}{\partial r^2} + \frac{1}{r} \frac{\partial C}{\partial r} \right). \tag{3.12}$$

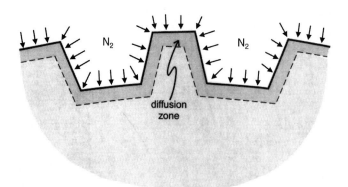

Figure 3.4 If the thickness of the diffusion zone is small with respect to the size of the gearwheel teeth, then we can approximate the diffuse flux as being planar.

While some problems do not conform to any of these symmetries, many of them either do or we can assume that they do to an adequate level of approximation. If they do not, then we must resort to numerical techniques, such as finite-element methods, to develop solutions. While this can be done, it certainly involves more work and requires access to appropriate software packages. Methods of this type are beyond the scope of this book and will not be discussed further.

Once we have established a geometry for the problem, we need to determine the boundary values. These are of two types, as noted earlier.

First, there are the initial conditions which define the state of the body at some well-defined time (usually designated to be zero time). In order to make problems accessible to analytical solutions we usually aim to define simple initial conditions. Most often we start by assuming the body is originally at equilibrium, in which case the composition within each phase is uniform. For example, the heat treatment of an alloy often begins with a high-temperature solutionizing step designed to put all of the alloy elements into solid solution in the matrix. If this process is carried out under the correct conditions of temperature and time an equilibrium will be established. The heat treatment then proceeds by exposing the alloy to lower temperature for various times during which second-phase particles are precipitated. We can predict what will happen by starting from the equilibrium state established during the initial solutionizing stage. If this is not the case then we need to find some simple analytical function (such as a sinusoidal distribution) which adequately describes the initial state.

Second, we arrive at the fun part – the boundary conditions. Fick's Second Law is *second* order in space, *i.e.* it contains the second derivative of the

Figure 3.5 If the diffusing species completely penetrates the gearwheel then a cylindrical geometry is more appropriate.

concentration with respect to position. This means that we need *two* independent boundary conditions. For diffusion, a boundary condition involves either the concentration or some function of it. The most common boundary conditions either invoke a fixed concentration or a fixed flux (*i.e.* a fixed value of $\partial C/\partial y$). Determining the appropriate boundary conditions is by far the most interesting (and the most challenging) part of setting up diffusional boundary value problems. However, we will soon find that most boundary conditions fall into a small number of categories. After all, how many types can there be if they involve either fixed concentration or fixed flux? Our task for the moment is to study the solutions that arise when the geometry, the initial conditions and the boundary conditions are known. We will devote much of the following chapters to a study of how we obtain these conditions in a wide range of applications.

Once we have assembled the components of a diffusion problem in terms of the geometry, initial conditions and boundary conditions, we need to determine what kind of questions we want to answer. If we are degassing an alloy we may want to know how long it will take until a certain fraction of the solute has been removed from the bulk. In this case we want information on the average concentration. In another case we may want to know when the diffusion fields due to adjacent particles begin to overlap, in which case we need information on the concentration at a specific location. For some problems we may want to turn this around and ask what the concentration profile will look like after a specified period of time. We may also want to know how long it will take until a steady state is established or how fast an interface moves through a solid. All of these questions can be answered, as we will see, by making use of standard diffusion solutions available in the literature. So let us look at some of these solutions and see what is available.

We will now begin to study the variety of solutions to Fick's Second Law. We will see that these solutions are of a different type depending on whether the times involved are 'long' or 'short'. We can use the short-time solutions only if the distance over which diffusion might occur is large with respect to a nominal diffusion distance $\bar{L} = \sqrt{Dt}$. We will see how this parameter arises shortly. But for now it provides a convenient 'road sign' in distinguishing when short-time solutions are applicable.

3.3 Solutions to Fick's Second Law at short time (far from equilibrium)

3.3.1 Plane initial source

We start by considering a simple problem in which a very thin layer of some element B is assumed to be sandwiched between two thick pieces of another element A. We can imagine making such a specimen by taking two pieces of A and plating the surface of one of these (*e.g.* by vapour deposition) with a thin layer of B. Let us suppose that this layer has a thickness δ. We then squeeze the two pieces together with the B layer in between, as shown in the lower part of Fig. 3.6, and assume a metallurgical bond is formed everywhere between A and B. The cross-sectional area of the piece is A. We now increase the temperature sufficiently so that significant diffusion can take place, and the B atoms begin to diffuse into A. If the diffusion distance is large with respect to δ, then we call this a 'plane initial source'. This is just the planar equivalent of a point source. The total amount of solute (B) in this thin initial layer is equal to the concentration of pure B, C^* (in moles per unit volume) times the volume of the layer, $A\delta$. We can therefore state the initial conditions (I.C.) for this problem to be:

$$\text{I.C.} \qquad C = C^*, \qquad\qquad -\delta/2 < y < +\delta/2$$

$$C = 0, \qquad\qquad \text{elsewhere.} \tag{3.13}$$

Remember that the concentration here refers to that of the B atoms. Now what about the boundary conditions? If the time is short then diffusion will not extend all of the way through the plates of material A. The boundary conditions (B.C.) are therefore given by the fact that the concentration of B far from the origin remains zero, *i.e.*

$$\text{B.C.} \qquad C = 0, \qquad\qquad y \to \pm\infty. \tag{3.14}$$

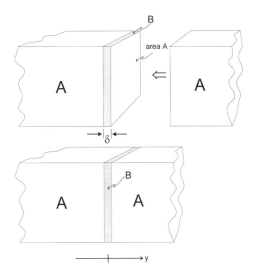

Figure 3.6 A plane initial source is made by plating a thin layer of solute B onto one end of a semi-infinite bar of material A, and butt welding it to another semi-infinite bar of A to form a metallurgical bond everywhere between A and B. This is assumed to occur without any diffusion of B away from the source.

Because the thickness of the initial layer is infinitesimal eq. (3.13) is not a very useful initial condition, unless we know precisely the thickness δ, which in most instances we will not. Thus this equation does not unambiguously define the conditions at the origin, and we need something else. We get that by invoking the principle of conservation of matter. This states that the total amount of solute remains constant and equal to the amount present initially. Thus

$$\int_{-\infty}^{\infty} C(y,t)A\,\mathrm{d}y = C^*A\delta \qquad (3.15)$$

defines the total amount of solute in the solid both initially and at all subsequent times. We now have a fully defined diffusion problem consisting of the governing equation – Fick's Second Law (given by eq. (3.4)) – and initial and boundary conditions (given by eqs. (3.13)–(3.15)). The solution to this problem is given by

$$C(y,t) = \frac{C^*\delta}{2\sqrt{\pi Dt}} \exp\left(-\frac{y^2}{4Dt}\right). \qquad (3.16)$$

Although we have not derived this solution, it is straightforward to show that this is the correct solution by substituting it into Fick's Second Law and testing the initial and boundary conditions.

There are two implicit assumptions used in developing eq. (3.16). First, the element B has to have sufficient solubility in A that no other phase forms. Second, we have assumed that the thickness δ of the original solute layer is small compared to the distance over which diffusion occurs. Since the exponential term is equal to unity at $y = 0$, the pre-exponential term defines the concentration of solute at the origin. It decays as $1/t^{1/2}$. The full solution is plotted in Fig. 3.7, along with the first and second derivatives of the concentration. The first derivative $\partial C/\partial y$ is proportional to the flux and so it tells us the regions in which the flux of solute is positive and negative. We see that the position of maximum flux moves outwards with time while the magnitude of the maximum flux decreases. From Fick's Second Law we know that the second spatial derivative of the concentration $\partial^2 C/\partial y^2$ is proportional to the time derivative $\partial C/\partial t$. Thus the negative region on this plot corresponds to the inner region in which solute concentration is decreasing with time. The width of this region increases with time.

Example 3.1: Rate of decay from a plane initial source

(a) Suppose that a thin 1 µm layer of a Ni alloy containing 10 wt% Cr is sandwiched between thick plates of pure Ni and that a metallurgical bond is formed between the Ni and the alloy. This sandwich is then heated at 1000°C.

How long must the heat treatment last in order to reduce the concentration at the centre to 1 wt% Cr?

(b) After this heat treatment, the Cr concentration profile is measured. At what distance from the centre will the Cr concentration fall to 0.1 wt%?

The diffusion coefficient for Cr in Ni is listed in Appendix B. It is given by $D = D_0 \exp(-Q/RT)$, where $D_0 = 1.1 \times 10^{-4} \, m^2/s$ and $Q = 272 \, kJ/mol$. From this we determine the diffusion coefficient at 1000°C to be $7.6 \times 10^{-16} \, m^2/s$.

(a) At the centre of the plate $y = 0$ and the exponential term equals unity at all time. Thus

$$C(0, t) = \frac{C^* \delta}{2\sqrt{\pi D t}}.$$

We can invert this to determine the time required:

$$t = \frac{1}{\pi D} \left(\frac{C^* \delta}{2 C(0, t)} \right)^2.$$

The initial and boundary conditions are given in the problem statement, in units of wt%. Strictly speaking, we should convert these to units of absolute concentration (*i.e.* moles or kg/m³). However, so long as we use normalized concentrations, such as C/C^*, the scaling factor cancels and the conversion is not necessary. By substitution we get the time needed for the heat treatment operation. It is 10,500 s, about 3 h.

(b) To determine the distance out from the centre to the position at which the concentration drops to a specific value C, we need to solve the

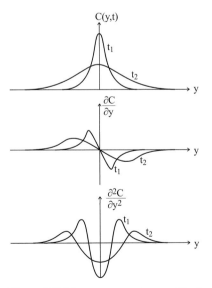

Figure 3.7 The concentration profile for a plane initial source is plotted (*top*) for two different times ($t_2 > t_1$). The flux, which is proportional to $-\partial C/\partial y$ (*middle*), is always positive for $x > 0$, and vice versa. The point of maximum flux moves outwards with time. $\partial C/\partial t$ is proportional to $\partial^2 C/\partial y^2$ (*bottom*).

full version of eq. (3.16) for y. The result is

$$y = \left[4Dt \ln\left(\frac{C^*\delta}{2\sqrt{\pi Dt}\, C} \right) \right]^{1/2}.$$

Since all of the terms are known we can substitute:

$$y = \Big[4 \times 7.6 \times 10^{-16} \times 1.05 \times 10^4$$

$$\times \ln\left(\frac{0.1 \times 10^{-6}}{2\sqrt{(\pi \times 7.6 \times 10^{-16} \times 1.05 \times 10^4)} \times 0.001} \right) \Big]^{1/2}$$

$$= 8.57 \times 10^{-6}\,\mathrm{m} = 8.57\,\mu\mathrm{m}$$

It is interesting to ask what would happen if we plated the solute onto the surface of a plate but did not embed it by attaching a second plate. This new configuration is shown in Fig. 3.8. In this case all of the solute must diffuse in one direction (the positive y-direction), whereas before the solute could diffuse in both directions. Therefore, for the same thickness of the solute layer δ, the concentration at any depth y from the surface should just be double what it was previously, *i.e.*

$$C(y, t) = \frac{C^*\delta}{\sqrt{\pi Dt}} \exp\left(-\frac{y^2}{4Dt} \right). \tag{3.17}$$

This method is often used to measure the diffusion coefficient by means of radioactive tracers. In this technique a thin layer of a radioactive isotope of some element is plated on to the surface and then allowed to diffuse into the material. After a sufficient time the sample is sectioned into several pieces and the amount of radioactivity of each piece is measured as a function of the distance of each section from the original surface. The radioactivity of a material is proportional to the number of radioactive atoms per unit volume. In other words, it is proportional to the concentration of the radioactive

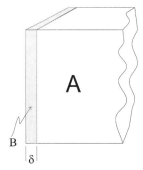

Figure 3.8 A schematic illustration of a surface source in which a thin layer of B is plated (without mixing) onto the surface of a semi-infinite bar of A.

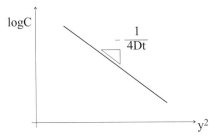

Figure 3.9 A plot of log (solute concentrations) vs depth squared has a slope $-1/(4Dt)$. This can be used to measure diffusivity. Since only the slope is needed, any quantity proportional to concentration (such as the radioactivity of an isotope) can also be used.

species. Therefore, if we plot the log of the radioactivity as a function of the depth squared, our equation tells us that the slope will be equal to $-1/(4Dt)$, as shown in Fig. 3.9. The nice thing about this is that we can measure D without knowing exactly how much solute we put on the surface or what constant of proportionality relates radioactivity with concentration. These factors will affect the intercept of the experimental line, but not its slope. Another useful thing about this method is that it becomes possible to measure the 'self-diffusion' coefficient, *i.e.* the diffusivity of an element in itself. Thus if we want to know how fast copper diffuses in pure copper we can plate a small amount of radioactive copper onto the surface of a plate of normal (non-radioactive) copper. While these two isotopes behave almost identically from a chemical viewpoint and have almost the same diffusion coefficient, radioactivity enables us to follow the penetration of the surface atoms into the bulk. Of course the same technique can also be used to measure solute diffusion. A large amount of diffusion data has been collected in this manner.[¶]

Example 3.2: Measurement of diffusion coefficient

A thin layer of radioactive copper is plated onto the end of a copper rod. The rod is then annealed for 20 hours, the specimen cut into thin sections, and the activity of each section measured. The following results are obtained.

Activity (counts/min mg of sample)	Position from end of rod (mm)
5012	0.1
3981	0.2
2512	0.3
1413	0.4
524.8	0.5

[¶] J. Askill, *Tracer Diffusion Data for Metals, Alloys, and Simple Oxides* (IFI/Plenum, New York, 1970)

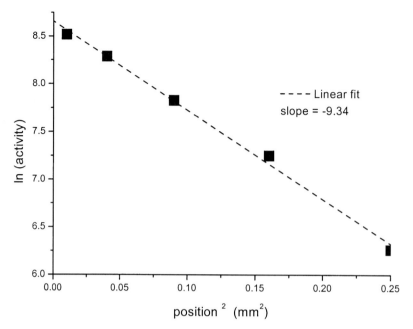

Figure 3.10 A plot of radioactivity vs position, plotted according to eq. (3.17) so that the slope equals $-1/(4Dt)$.

Plot this data in an appropriate way and determine the value of the self-diffusion coefficient of copper at this temperature. State any assumptions you are making.

We plot the log of the activity as a function of (position)^2, as shown in Fig. 3.10. A linear fit to this curve gives a slope of -9.34. From eq. (3.17), this slope equals $-1/(4Dt)$. Since $t = 20\,\text{h} = 7.2 \times 10^4\,\text{s}$, we find that

$$D = \frac{1}{4(9.34)(7.2 \times 10^4)} = 3.72 \times 10^{-7}\,\text{m}^2/\text{s}$$

The assumptions implicit in this analysis are that the original layer of radioactive copper on the surface was sharp and much thinner than the diffusion distances involved (*i.e.* $\delta \ll 5\,\text{mm}$). Note that in order to obtain the diffusion coefficient in the desired unit of m^2/s, we needed to convert the position data from the units given (mm) to units of m.

3.3.2 Diffusion from a distributed source

We will now consider what happens if we attach a piece of pure A to another piece of A that has some B dissolved uniformly within it. For example we might take a piece of pure copper and another piece with copper containing 5% zinc. Let us suppose that the initial concentration of B in the alloyed region is denoted as C^*. As before, we are interested in the solution at relatively short times, which means that diffusion does not extend through either piece. This situation is illustrated in Fig. 3.11. We will assume that we are able to butt weld these two pieces together to form a metallurgical bond without any intermixing taking place. We can therefore write the initial conditions as:

I.C. $C = 0,$ $y < 0$

         $C = C^*$ $y \geq 0.$ (3.18)

As diffusion proceeds the B atoms will move down the concentration gradient near the origin, and we would like to know how fast this will occur. In establishing the boundary conditions we again rely on the concept of short time, which means that far from the interface between the two pieces the concentrations remain unchanged. Therefore,

B.C. $C = 0,$ $y \to -\infty$

         $C = C^*,$ $y \to +\infty.$ (3.19)

We now have a well-defined problem and so we can develop a solution based on the mathematics of partial differential equations. However, in this case at least, it is easier to build up the solution from that for the plane initial source. We do this by dividing the piece for all positive y into thin slabs, each of thickness Δa. Each slab sits at a distance $a_i = i \cdot \Delta a$ from the origin as shown in Fig. 3.11. Each slab initially contains an amount of solute equal to $C^* \cdot \Delta a \cdot A$, where A is the cross-sectional area of the slab. Over time the

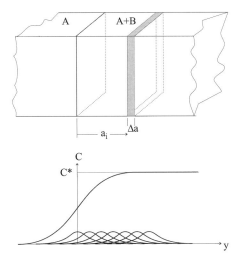

Figure 3.11 The solution for a distributed source can be derived by dividing the region containing solute into an infinite series of plane sources, each of width Δa. When these are summed, the sigmoidal solution illustrated here is obtained. This solution is described mathematically by the error function.

solute will diffuse away from this slab. If the slab is thin with respect to the average diffusion distance then the concentration profile from this one thin slab is just equal to that for the plane initial source (eq. 3.16), *i.e.*

$$C(y,t) = \frac{C^*\Delta a}{2\sqrt{\pi Dt}} \exp\left[-\frac{(y-a_i)^2}{4Dt}\right]. \tag{3.20}$$

Note, however, that the origin has been shifted to the centre of each slab at $y = a_i$.

We now need to think about how the solutions for each of these thin slabs can be combined to yield the total concentration profile. In order to do this we invoke the *principle of superposition*. This says that the solution to a set of diffusion problems with the same governing equation, the same geometry and the same boundary conditions but with different initial conditions can be added. The resulting equation gives the solution to the problem for which the initial condition is the sum of the initial conditions of the component problems. This works because it obeys the principle of conservation of mass. In other words, by summing the initial conditions we are taking care of all the solute initially present in the solid. Since the individual solutions conserve mass, as we have already seen, the summed solution must also be mass conservative.

In this case we have a set of solutions (given by eq. (3.20)) in which the initial conditions are given by

> I.C. $C = C^*$, $a_i - \frac{1}{2}\Delta a < y < a_i + \frac{1}{2}\Delta a$
>
> $C = 0$,   elsewhere.

The sum of these initial conditions for i running from zero to infinity is just the initial condition we are seeking, eq. (3.18). Thus the solution to our problem is given by summing the plane initial source solution, *i.e.*

$$C(y,t) = \sum_{i=0}^{\infty} \frac{C^*\Delta a}{2\sqrt{\pi Dt}} \exp\left[-\frac{(y-a_i)^2}{4Dt}\right]. \tag{3.21}$$

If we allow the width Δa of the slabs to become infinitesimally small then the summation can be replaced by an integral:

$$C(y,t) = \frac{C^*}{2\sqrt{\pi Dt}} \int_0^{\infty} \exp\left[-\frac{(y-a)^2}{4Dt}\right] da. \tag{3.22}$$

The solution to this integral is made easier if we make a substitution, namely

$$\eta = \frac{y-a}{2\sqrt{Dt}}. \tag{3.23}$$

The integral then becomes

$$C(y,t) = \frac{C^*}{\sqrt{\pi}} \int_{-\infty}^{y/2\sqrt{Dt}} \exp(-\eta^2)\, d\eta. \tag{3.24}$$

The solution of this integral is a special function known as the 'error function'. Values for this function are available in tables (see Table 3.2). For those of you not familiar with the error function and its derivation, refer to the box below. The integral in eq. (3.24) can be written in two parts as

$$C(y, t) = \frac{C^*}{\sqrt{\pi}} \left[\int_{-\infty}^{0} \exp(-\eta^2) \, d\eta + \int_{0}^{y/2\sqrt{Dt}} \exp(-\eta^2) \, d\eta \right],$$

the solution for which is

$$C(y, t) = C^* \left[\frac{1}{2} + \frac{1}{2} \operatorname{erf}\left(\frac{y}{2\sqrt{Dt}} \right) \right]$$

$$= \frac{C^*}{2} \left[1 + \operatorname{erf}\left(\frac{y}{2\sqrt{Dt}} \right) \right]. \tag{3.25}$$

It is easy to show that this solution matches the boundary conditions through substitution. It is also worth noting that the concentration at the original interface between the two plates is a constant, equal to $\frac{1}{2}C^*$. But we must remember that this is only the case so long as the diffusion coefficient remains a constant. We will return to this later and consider what happens if this is not the case.

Table 3.2 Table of error function values.

z	erf (z)	z	erf (z)
0	0	0.85	0.7707
0.025	0.0282	0.90	0.7970
0.05	0.0564	0.95	0.8209
0.10	0.1125	1.0	0.8427
0.15	0.1680	1.1	0.8802
0.20	0.2227	1.2	0.9103
0.25	0.2763	1.3	0.9340
0.30	0.3286	1.4	0.9523
0.35	0.3794	1.5	0.9661
0.40	0.4284	1.6	0.9763
0.45	0.4755	1.7	0.9838
0.50	0.5205	1.8	0.9891
0.55	0.5633	1.9	0.9928
0.60	0.6039	2.0	0.9953
0.65	0.6420	2.2	0.9981
0.70	0.6778	2.4	0.9993
0.75	0.7112	2.6	0.9998
0.80	0.7421	2.8	0.9999

Aside: The Mathematics of the Error Function

Consider a distribution function with the form

$$\varphi(n) = \frac{1}{\sqrt{\pi}} e^{-n^2}$$

It can be shown that this function, integrated from $-\infty$ to $+\infty$, is equal to unity, *i.e.*

$$\frac{1}{\sqrt{\pi}} \int_{-\infty}^{+\infty} e^{-n^2} \, dn = 1$$

We define the error function as

$$\operatorname{erf}(z) \equiv \frac{1}{\sqrt{\pi}} \int_{-z}^{+z} e^{-n^2} \, dn$$

$$= \frac{2}{\sqrt{\pi}} \int_{0}^{+z} e^{-n^2} \, dn$$

It is readily seen that the error function is antisymmetric, *i.e.* $\operatorname{erf}(z) = -\operatorname{erf}(-z)$. Moreover, $\operatorname{erf}(0) = 0$, while $\operatorname{erf}(\infty) = 1$. It is useful to also define the error function compliment as $\operatorname{erfc}(z) = 1 - \operatorname{erf}(z)$. While no exact analytical solution exists to the error function, an approximate solution is available in terms of an infinite series:

$$\operatorname{erfc}(z) = \frac{1}{\sqrt{\pi}} e^{-z^2} \left(\frac{1}{z} - \frac{1}{2z^2} + \frac{1 \cdot 3}{2^2} \frac{1}{z^5} - \cdots \right)$$

Since this is an alternating series, the error introduced by truncating the series after n terms is about as great as the $(n+1)$th term. For large z, an adequate approximation is found by taking only the first term, *i.e.*

$$\operatorname{erfc}(z) \approx \frac{1}{\sqrt{\pi}} \frac{e^{-z^2}}{z}$$

3.3.3 Diffusion from a surface at fixed concentration

We can apply this result directly to the problem of diffusion into a solid from a surface which has a fixed solute concentration, say C_s, as shown in Fig. 3.12. This is a very common situation since a surface will often be in local equilibrium with the surrounding gas phase, which leads to a fixed concentration at the surface. If the solid is originally solute free then the initial condition is simply

$$\text{I.C.} \qquad C = 0, \qquad\qquad y > 0 \tag{3.26}$$

The boundary conditions are given (for short time) by

$$\text{B.C.}\quad \begin{aligned} C &= C_s, & y &= 0 \\ C &= 0, & y &\to \infty \end{aligned}$$

(3.27)

We can obtain the solution to this problem directly from that for a distributed source, eq. (3.25), by noting that in that case

$$C(y, t) = \tfrac{1}{2}C^*, \quad \text{at } y = 0.$$

(3.28)

To use this solution we have merely to replace $\tfrac{1}{2}C^*$ by C_s and invert the direction of flow. The last requirement is necessary since in the previous section the solute flowed in the negative y-direction and we have configured the current problem so that diffusion is in the positive y-direction. This inversion is obtained by noting that $\operatorname{erf}(-z) = -\operatorname{erf}(+z)$. Therefore the solution is

$$C(y, t) = C_s\left[1 - \operatorname{erf}\left(\frac{y}{2\sqrt{Dt}}\right)\right] = C_s\operatorname{erfc}\left(\frac{y}{2\sqrt{Dt}}\right).$$

(3.29)

As we see in Fig. 3.12, the concentration remains fixed at the surface equal to C_s and decays gradually towards zero into the interior of the solid. The distance over which diffusion occurs increases with time.

How would the solution to this problem be different if the initial concentration were some constant concentration other than zero, say $C = C_0$? Because of the superposition principle introduced in the previous section, the answer is trivial. The concentration is simply increased everywhere by this amount, *i.e.*

$$C(y, t) = C_0 + (C_s - C_0)\operatorname{erfc}\left(\frac{y}{2\sqrt{Dt}}\right).$$

(3.30)

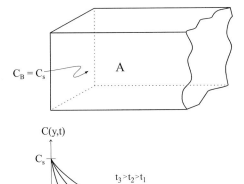

Figure 3.12 The solution for a surface source, in which the solute concentration at the surface is fixed equal to C_s, can be derived simply from that for a distributed source.

Thus when $C_s = C_0$ there is no diffusion, as we expect. Moreover, if $C_s < C_0$, solute diffuses *out* of the solid.

Example 3.3: Doping of semiconductors – I

A 2 mm thick silicon wafer is to be doped with antimony (Sb) in order to create a p-type region. You do this by passing a $SbCl_3/H_2$ gas mixture over the surface of the wafer at $1200°C$, which fixes the surface Sb concentration at $10^{23}/m^3$. Suppose that you want the donor density (which is just another term for the Sb concentration) to be greater than or equal to $3 \times 10^{22}/m^3$, over a depth of $1 \mu m$ below the surface. Determine how long the wafer should be exposed to this atmosphere in order to achieve this.

The initial and boundary conditions can be easily obtained, if we assume that the Si wafer is initially free of Sb.

I.C. $C = 0$, $y \geq 0$, $t = 0$

B.C. $C = C_s$, $y = 0$

$C = 0$, $y \to \infty$,

where C_s is given as $10^{23}/m^3$. From the development we have just completed, we know that the concentration profile due to this set of initial and boundary conditions is given by

$$C = C_s \operatorname{erfc}\left(\frac{y}{2\sqrt{Dt}}\right).$$

We wish to find the time t, for which $C = 3 \times 10^{22}/m^3$ at $y = 1$ μm. Thus

$$\operatorname{erf}\left(\frac{y}{2\sqrt{Dt}}\right) = 1 - \operatorname{erfc}\left(\frac{y}{2\sqrt{Dt}}\right) = 1 - \frac{C}{C_s} = 0.7.$$

From the error function table, this condition is met when $y/2\sqrt{Dt} \approx 0.73$. We must now evaluate the diffusion coefficient at 1200°C. It is

$$D = D_0 \exp\left(-\frac{Q}{RT}\right) = 1.3 \times 10^{-3} \exp\left(-\frac{383,000}{8.314 \times 1473}\right)$$

$$= 3.4 \times 10^{-17} \, m^2/s.$$

By substitution we find that the required time is

$$t = \frac{1}{D}\left[\frac{y}{2 \times 0.73}\right]^2 = 1.38 \times 10^4 \, s$$

$$\approx 3.8 \, h.$$

Now let us take this a little further and ask how much solute has diffused into the solid after a certain period of time. We can do this by integrating eq. (3.30) in order to determine the time-averaged flux or the total solute absorbed or removed at the interface. The average flux over an exposure time t_e is equal to

$$\bar{J} = \frac{1}{t_e} \int_0^{t_e} J|_{y=0} \, dt. \tag{3.31}$$

The surface flux

$$J|_{y=0} = -D\frac{\partial C}{\partial y}\bigg|_{y=0} = -D(C_s - C_0)\frac{\partial}{\partial y}\left[\text{erfc}\left(\frac{y}{2\sqrt{Dt}}\right)\right]\bigg|_{y=0}. \tag{3.32}$$

The slope of the error function can be obtained from its definition

$$\frac{\partial}{\partial y}\left[\text{erfc}\left(\frac{y}{2\sqrt{Dt}}\right)\right]\bigg|_{y=0} = -\frac{\partial\,\text{erf}(z)}{\partial z}\bigg|_{z=0}\cdot\frac{1}{2\sqrt{Dt}}$$

$$= -\frac{2}{\sqrt{\pi}}\cdot\frac{1}{2\sqrt{Dt}} = -\frac{1}{\sqrt{\pi Dt}}. \tag{3.33}$$

Thus,

$$J|_{y=0} = +(C_s - C_0)\sqrt{\frac{D}{\pi t}}. \tag{3.34}$$

The time-averaged flux is therefore given by substituting this equation back into eq. (3.31):

$$\bar{J} = \frac{C_s - C_0}{t_e}\sqrt{\frac{D}{\pi}}\int_0^{t_e}\frac{dt}{\sqrt{t}} = 2(C_s - C_0)\sqrt{\frac{D}{\pi t_e}}. \tag{3.35}$$

Example 3.4: Doping of semiconductors – II

Suppose that in the doping process described in Example 3.3 the process had been accidentally allowed to run for 5 hours instead of the 3.8 hours that was needed.
(a) By how much would the Sb level be increased at a depth of 1 μm below the surface?
(b) By how much would the total amount of Sb absorbed in the wafer have been increased?

(a) The Sb concentration at a given depth is given by eq. (3.29). We first need to evaluate $y/2\sqrt{Dt}$. At a time of $t = 5\,h = 1.8 \times 10^4\,$s, this is equal to 0.64. From the error function table, $\text{erfc}(y/2\sqrt{Dt})$ is about $1 - 0.63 = 0.37$. By substitution into eq. (3.29) we therefore find the concentration to be $3.7 \times 10^{22}/\text{m}^3$ at $y = 1\,\mu$m. In other words, the increased time has increased the concentration at this depth by about 23%.

(b) To answer this question we need to calculate the total amount of Sb absorbed per unit area after 3.8 and 5 hours and compare these. This amount, call it q, is equal to the time-averaged flux times the exposure time, *i.e.*

$$q = 2(C_s - C_0)\sqrt{\frac{Dt_e}{\pi}}$$

This is equal to $7.8 \times 10^{16}/\text{m}^2$ after 3.8 hours and $8.9 \times 10^{16}/\text{m}^2$ after 5 hours, an increase of 14%.

3.3.4   Diffusion between two solids with different initial solute concentrations

Another problem of this type involves putting two solids together, both of which contain solute but of different composition. Clearly, diffusion will occur so as to produce a material with a uniform solute concentration. This is rather like the original problem of diffusion from a distributed source, illustrated in Fig. 3.11. We merely need to realize that diffusion is driven by the *difference* in concentration between the two pieces. Thus if the initial conditions are

$$
\text{I.C.} \qquad
\begin{aligned}
C &= C_1, & y &< 0 \\
C &= C_2, & y &\geq 0,
\end{aligned}
\tag{3.36}
$$

then the diffusive flux is governed by $(C_2 - C_1)$. At short times the boundary conditions are again determined by the unchanged concentration far from the interface. By comparison with the result for the distributed solution (eq. (3.25)) we can write

$$
C(y, t) = C_1 + \frac{C_2 - C_1}{2} \left[1 + \operatorname{erf}\left(\frac{y}{2\sqrt{Dt}} \right) \right].
\tag{3.37}
$$

If we set $C_0 = \frac{1}{2}(C_1 + C_2)$, then

$$
\frac{C(y, t) - C_1}{C_0 - C_1} = 1 + \operatorname{erf}\left(\frac{y}{2\sqrt{Dt}} \right).
\tag{3.38}
$$

In this form it is clear that at the original joint between the two pieces, *i.e.* at $y = 0$, $C(y, t)$ equals C_0 at all (short) times.

3.4   The nominal diffusion distance

In all of the solutions we have just developed there is one parameter that appears repeatedly. This is $\sqrt{Dt}$. It has units of length, and must therefore be related to some characteristic length in the material. To understand what this might be we can consider again the solution for diffusion into a solid from a surface with a fixed surface concentration C_s. In order to simplify things, and without compromising the generality of the analysis, we can assume that the initial concentration C_0, is zero. Therefore the solution takes the form

$$
C(y, t) = C_s \operatorname{erfc}\left(\frac{y}{2\sqrt{Dt}} \right).
\tag{3.39}
$$

This is plotted in Fig. 3.13. If we examine tables of error function values (see Table 3.2) we find that $\operatorname{erfc}(z) \approx 0.5$ when $z = 0.5$. In other words, the concentrations drops to half of its surface value over a distance equal to $\sqrt{Dt}$.

Moreover, the concentration drops further to almost 0.1 C_s over a distance of $2\sqrt{Dt}$ from the surface. Thus most of the diffusion is occurring within this distance from the surface. We therefore call $\sqrt{Dt}$ the 'nominal diffusion distance', $\bar{L}$.

This parameter can be used in a variety of ways to guide the development of solutions to diffusion problems. Since $\sqrt{Dt}$ is approximately the distance over which diffusion can occur in a time t, we can use it to estimate when the short-time solutions we have just developed are valid. In these solutions we have repeatedly assumed that the concentration does not differ from its original value far from an interface. We can now be more concrete. The far side of a solid will not be influenced by diffusional processes which occur at a distance that is large with respect to $\sqrt{Dt}$.

The 'nominal diffusion distance' should also be determined when you first start to tackle a problem. This gives an order of magnitude estimate of the distance or time involved in a problem. You can use this idea in a variety of ways. If you wish to perform an operation at a given temperature (thus fixing the diffusion coefficient D), and the operation should be complete within a given time period t, then you can estimate the nominal distance over which diffusion can occur, namely $\bar{L} = \sqrt{Dt}$. If, on the other hand, you know from the microstructure of the material the scale $\bar{L}$ over which diffusion must occur in order to achieve a desired microstructural change, then you can estimate the

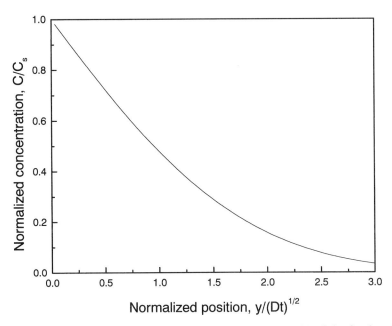

Figure 3.13 The surface concentration drops to about 50% of the fixed surface concentration over a distance equal to $\sqrt{Dt}$. We call this distance the 'nominal diffusion distance'.

time required at the temperature of interest, $t = \bar{L}^2/D$. If you find that the time is much too long then you can estimate what temperature you would need to go to in order to reduce the time involved to a reasonable level by calculating the diffusion coefficient you would need, $D = \bar{L}^2/t$. It is important to remember however, that this simple relationship gives only a first-order estimate.

Example 3.5: Nominal diffusion distance

When a nickel-base superalloy is exposed to steam at high temperatures, hydrogen can penetrate the material to significant depths. Estimate this by determining the average penetration depth of hydrogen into pure Ni at 800°C, for 1, 10 and 100 hours. Note that when hydrogen diffuses into a metal such as nickel it dissociates into individual H atoms.

We can estimate the approximate penetrate depth as $\sqrt{Dt}$. For H in Ni, using data in Appendix B, the diffusion coefficient at 800°C is

$$D = 4.8 \times 10^{-9} \exp\left(\frac{39,300}{8.314 \times 1073}\right) = 5.86 \times 10^{-11} \, \mathrm{m^2/s}$$

From this we find that $\bar{L} = \sqrt{Dt}$ is equal to 0.48, 1.45 and 4.8 mm for 1, 10 and 100 hours exposure respectively.

Another application of the nominal diffusion distance is in equating the amount of diffusion that can occur at different temperatures. Recall that the diffusion coefficient depends on temperature according to the following relationship:

$$D = D_0 \exp\left(-\frac{Q}{RT}\right). \tag{3.40}$$

Suppose that we wish to know how much faster a process will be complete at a temperature T_2 than at a lower temperature T_1. Since the distance over which diffusion must occur is the same then

$$\bar{L} = \sqrt{D_1 t_1} = \sqrt{D_2 t_2}, \tag{3.41}$$

and so the relative amount of time required is related by

$$\frac{t_2}{t_1} = \frac{D_1}{D_2} = \exp\left[-\frac{Q}{R}\left(\frac{1}{T_1} - \frac{1}{T_2}\right)\right]. \tag{3.42}$$

We must use this relationship with some care, however, since it assumes that the same diffusional process controls the kinetics at both temperatures. It is always possible that the controlling mechanism, and thus the relevant diffusion process, may change with temperature. In this case the calculation just described will be in error.

We can extend this concept to deal with the effect of continuous heating and cooling. This is important for example in the batch annealing of large coils of steel, or in the firing of a ceramic component in which low heating and cooling rates are required in order to avoid cracking due to thermal shock. Consider the time–temperature cycle illustrated in Fig. 3.14, in which a piece of material is heated slowly to the nominal heat treatment temperature T_a. Even though the diffusion coefficient is quite sensitive to temperature, some diffusion will occur at temperatures below T_a. We would like to know what would be the equivalent time at T_a due to the heating and cooling period. We can divide the heating cycle into N segments, each of time Δt. For each segment i the temperature is T_i. The equivalent time if spent at temperature T_a is equal to

$$\Delta t_a = \Delta t \cdot D(T_i)/D(T_a).$$

If we sum these, then the equivalent time during heating if spent at temperature T_a is

$$t_e = \sum_{i=1}^{N}(\Delta t_a)_i = \Delta t \sum_{i=1}^{N} \exp\left[-\frac{Q}{R}\left(\frac{1}{T_i} - \frac{1}{T_a}\right)\right]. \tag{3.43}$$

If we allow the size of the time interval Δt to be reduced towards zero then the summation can be replaced by an integral over the total time of heating t_h, giving

$$t_e = \int_0^{t_h} \exp\left[-\frac{Q}{R}\left(\frac{1}{T(t)} - \frac{1}{T_a}\right)\right] dt. \tag{3.44}$$

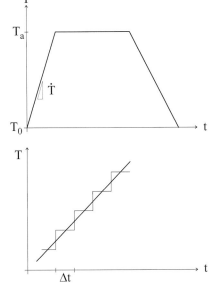

Figure 3.14 A schematic illustration of a time–temperature cycle in which a solid is heated slowly to a temperature T_a, then held there fore some time, and slowly cooled. the equivalent diffusion time at T_a can be estimated using the nominal diffusion distance.

Moreover, using this equation, any form of heating (or cooling) curve can be treated, not just the linear one illustrated in the figure.

3.5   Solution methods for Fick's Second Law at longer time (near equilibrium)

When the diffusion distance approaches $\sqrt{Dt}$, the short-time approximation is no longer valid. As a rule of thumb, this breaks down when $\sqrt{Dt}$ exceeds about 20% of the distance involved (*i.e.* $Dt/L^2 \approx 0.04$), although this will depend somewhat on geometry. In this section we will look at some of the same problems we considered previously but using different boundary conditions valid at longer times as equilibrium is approached.

3.5.1   Diffusion in a slab from a plane initial source

We will start, as before (see Section 3.3.1), by considering what happens when a thin layer, thickness δ, of some solute B is sandwiched between two thicker, but finite, pieces of another element A. The thickness of the B layer is assumed to be much less than that of the slab $2L$, as illustrated in Fig. 3.15. The element B is also assumed to be sufficiently soluble in A that no intermediate phases form. The initial conditions for this problem are exactly as before:

$$\text{I.C.} \qquad C = C^*, \qquad -\delta/2 < y < +\delta/2$$
$$C = 0, \qquad \text{elsewhere.}$$

Once the slab is heated to a temperature sufficient to allow diffusion to occur at a reasonable rate the solute will begin to flow away from the centre of the slab down the concentration gradient. Initially, the solution will have the form developed in the previous section. However, after some time, diffusion extends to the ends of the slab. Unless the solute is volatile it is trapped within the solid. Thus there is no flux of solute across the surface. Since the flux must be continuous everywhere except when a new phase is forming the flux must go to zero at the end of the slab. Thus the boundary conditions become

$$\text{B.C.} \qquad \partial C/\partial y = 0 \qquad y = \pm L$$

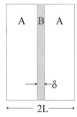

Figure 3.15 A schematic illustration of the plane initial source in a slab of finite width $2L$.

We can also invoke the condition of conservation of matter which we used previously. This requires that the total amount of solute within the slab is fixed. Therefore (similar to eq. (3.12))

$$\int_{-L}^{+L} C(y,t)\,dy = C^*\delta. \tag{3.45}$$

There are a number of ways in which we can obtain a solution to this problem. For example, we can use separation of variables to solve Fick's Second Law subject to the boundary conditions just described (see Appendix A). However, there is another interesting approach we can use to obtain solutions to a wide variety of problems of this type. We will discuss this next.

3.5.2 Use of image sources

The concept of image sources is used widely in various branches of physics and engineering in order to determine the effect of surfaces or interfaces on properties. What is done in each case is to perform a simple thought experiment in which we remove the surface and instead treat an infinite body, but with an imaginary source at some location beyond the surface which creates an equivalent effect. This imaginary source produces a mirror image of that in the body and leads to conditions along the 'interface' (which we have now removed) which mimic the initial conditions of the problem.

This concept is best understood by means of an example, so let us consider the diffusion problem at hand. Suppose that we replace the slab of width $2L$ with an infinitely wide slab. We want to find a solution to Fick's Second Law which still satisfies the conditions that $\partial C/\partial y = 0$ at $y = \pm L$. We also want to ensure that the amount of solute is fixed in the region between $-L$ and $+L$, as given by eq. (3.45). If we do this over the region of interest we will have satisfied all of the imposed conditions and the solution we obtain will be the same as that for the original problem. One way to manage this is to place a series of plane initial sources, all identical to that at the origin, with a periodic spacing of $2L$. This is illustrated in Fig. 3.16.

We then add the solution due to each source, based on the principle of superposition discussed earlier. We can see that at $y = +L$, the solution coming from the source at $y = +2L$ is of equal magnitude, but the slope is of opposite sign, to that from the source at the origin. Therefore the sum of these produces a solution with zero slope. Moreover this solution is equivalent to reflecting the original solution at $y = L$ so that the level of solute is conserved with the inner region of interest. The same occurs at $y = -L$ due to the source at $y = -2L$. Now this would be enough if the solution decayed quickly. But if it doesn't there is still a finite concentration at $y = -L$ due to the source at $y = 2L$, a distance $3L$ away. The 'image' of this source acting at $y = -L$ is a plane initial source at $y = -4L$.

Therefore as we carry this further it becomes clear that an infinite series of sources is required. Each additional source accounts for any of the concentration profile that 'escapes' through the surface and thus ensures that the conservation condition is satisfied. Another way of thinking about this is to imagine that the single solution to a source at $y = 0$ in an infinite solid is 'reflected' back into the solid at $y = L$ and $-L$. With this explanation we can now write down the solution. It is

$$C(y, t) = \sum_{n=-\infty}^{\infty} \frac{C^*\delta}{2\sqrt{\pi Dt}} \exp\left[-\frac{(y - 2nL)^2}{4Dt} \right]. \tag{3.46}$$

Comparing this with the solution for a single plane initial source we see that the only difference is the $(y - 2nL)$ term which shifts the origin of each source by $2L$. It is straightforward to substitute this solution back into Fick's Second Law and convince yourself that it does indeed match both the governing equation and the boundary conditions.

Now let us consider a slightly different situation in which the solute, upon reaching the surface, can evaporate or react with the vapour phase. An obvious example is the diffusion of nitrogen in a metal surrounded by a gas

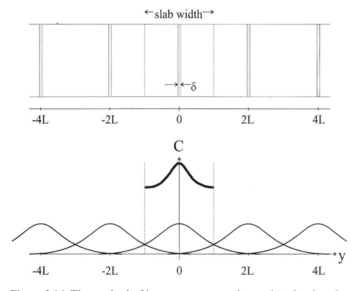

Figure 3.16 The method of image sources can be used to develop the solution for a plane initial source in a finite slab, from that for an infinite slab. the surface of the finite slab is assumed to be impermeable in this case. The thin curves are a series of solutions for the problem in an infinite body, with the origin of each solution displaced by $2L$. The bold curve is the sum of all these solutions within the region $-L < y < L$.

with a certain partial pressure of nitrogen. In this case the solute concentration is fixed by local equilibrium at the surface. We will consider the situation for which the surface concentration is essentially zero. How can we use the concept of image sources to help us here? Well, if we wish to make the solution for an infinite solid go to zero at $y = L$ we need to **subtract** a solution for an image source which has the correct value at that location. To do this we set up an image source at $y = 2L$, but with opposite sign (see Fig. 3.17).

This may seem rather strange at first. After all what does this mean, having a source with a negative concentration? What does 'anti-matter' have to do with diffusion? But just remember that this is merely a convenient way of finding a solution to a mathematical problem, and we will check it out against the boundary conditions later on. As before, we need to take care of the tails in the concentration profile by adding further sources. But now the sign of each source is opposite to its neighbour. To see why this is the case, consider the concentration due to the negative source at $y = 2L$. This results in a small negative concentration at $y = -L$. The image for this, which makes the concentration go to zero here, is a positive source at $y = -4L$ (see Fig. 3.17). The final solution is therefore

$$C(y, t) = \sum_{n=-\infty}^{\infty} (-1)^n \frac{C^* \delta}{2\sqrt{\pi D t}} \exp\left[-\frac{(y - 2nL)^2}{4Dt} \right]. \tag{3.47}$$

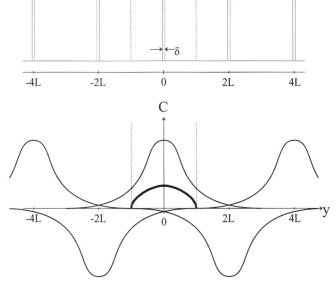

Figure 3.17 The image source method can also be used when the surface has a fixed concentration. The thin curves are a series of solutions for the problem in an infinite body, with the origin of each solution displaced by 2L. The bold curve is the sum of all these solutions within the region $-L < y < L$.

The only difference from the previous solution is the $(-1)^n$ term, which forces the sign of each source to alternate. As before, we can substitute this solution back into Fick's Second Law and show that this solution does indeed match both the governing equation and the boundary conditions.

This method can be used to obtain solutions to a wide range of problems. For example consider the situation illustrated in Fig. 3.18, in which a piece of an (A–B) alloy of width $2h$, is sandwiched between two finite slabs of pure A. This is rather like the plane initial source except that the thickness of the inner layer is now finite. Suppose that the solute is confined within the solid so that the flux is zero at either end of the piece ($y = \pm L$). This can again be replaced by an infinite bar with sources of width $2h$ spaced a distance $2L$ apart. The solution for the single source in an infinite bar is given by two error function solutions and that for the finite slab is therefore given by an infinite series of these. The concentration profile is therefore given by

$$C(y,t) = \frac{C^*}{2} \sum_{n=-\infty}^{\infty} \left[\mathrm{erf}\left(\frac{h + 2nL - y}{2\sqrt{Dt}}\right) + \mathrm{erf}\left(\frac{h - 2nL + y}{2\sqrt{Dt}}\right) \right]. \qquad (3.48)$$

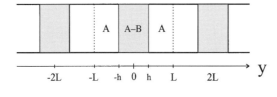

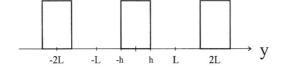

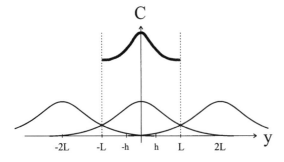

Figure 3.18 This illustrates the image source method applied to a distributed source. The thin curves are a series of solutions for the problem in an infinite body, with the origin of each solution displaced by $2L$. The bold curve is the sum of all these solutions within the region $-L < y < L$.

3.5.3   General solutions to Fick's Second Law

While the image source method can be helpful in finding a number of solutions to problems for which we have a solution to the infinite boundary problem, many of the solutions we need must be obtained through the normal process of developing mathematical solutions to Fick's Second Law, subject to the given initial boundary conditions. Two methods are available, both of which involve standard procedures for the solution of partial differential equations. The first involves the *separation of variables* technique. This method leads, for planar geometries, to solutions which are made up of infinite series of trigonometric functions. The second method involves the use of Laplace transforms. This leads to solutions made up of infinite series of error functions. Both of these solution methods have their uses. This is because the error-function solutions converge rapidly (that is, you only need a few terms) at relatively short times, while the trigonometric solutions converge more rapidly at longer times. The mathematics behind these solution methods are discussed in considerable detail in standard textbooks on partial differential equations.[¶] There is also a brief but practical summary of these methods in Appendix A.

The separation of variables method leads to a single general solution of Fick's Second Law. For planar symmetry it is

$$C(y, t) = \sum_{n=0}^{\infty} [A_n \sin(\lambda_n y) + B_n \cos(\lambda_n y)] \cdot \exp(-\lambda_n^2 Dt). \tag{3.49}$$

The constants A_n, B_n and the eigenvalues λ_n are determined by applying the initial and boundary conditions. An example is given in Appendix A.

One rather simple application of this general solution is for homogenization. Suppose that a material develops a non-uniform distribution of some solute species. This can easily happen for example during solidification, in which the first material to solidify is generally solute poor while the last is solute rich. In conventional casting processes solidification involves the growth of dendritic needles into the liquid with the last metal to solidify filling in between these dendrites. Thus the solute concentration is periodic throughout the casting on a scale which depends on the average spacing of the primary and secondary dendrite arms. This variability in the solute content can persist through subsequent processing steps such as hot and cold rolling. An example is shown in Fig. 3.19.

Let us suppose that the initial solute concentration profile can be described by a simple sinusoidal profile of the form

$$\text{I.C.} \qquad C = C_0 + \alpha \cos\left(\frac{\pi y}{L}\right). \tag{3.50}$$

[¶] See, for example, E. Kreyszig, *Advanced Engineering Mathematics*, 7th edition (Wiley, New York, 1993). You should note however that in mathematics texts Fick's Second Law is generally referred to as the 'heat equation'.

Here C_0 is the average solute concentration, while α and $2L$ are the amplitude and periodicity respectively, of the non-uniformity. This is illustrated in Fig. 3.20. Equation (3.50) represents the initial condition for this problem. The boundary conditions are defined by the periodicity of the problem, which requires that

B.C. $\partial C / \partial y = 0$, $y = \pm L$

In order to solve this problem, it useful to make a simple transformation by letting $\xi = C - C_0$. The addition of a fixed concentration does not change the general solution to Fick's Second Law and the initial and boundary conditions

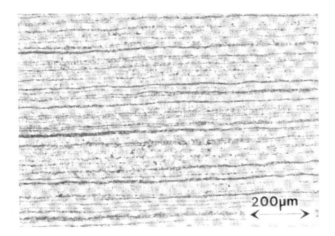

Figure 3.19 An example of a banded microstructure in a C–Mn steel following rolling. The variation in microstructure due to a variation in solute content is clearly visible, with a periodicity of about 20 μm (A. From and R. Sandstrom, Assessment of banding of steels using advanced image analysis, Mater. Charact., **41**, 11 (1998).

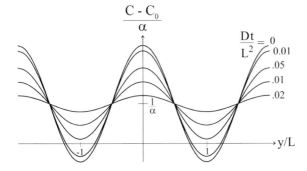

Figure 3.20 The normalized concentration $(C - C_0)/\alpha$ plotted as a function of distance y/L for a range of normalized times Dt/L^2, during homogenization of an initially sigmoidal distribution.

are altered only slightly, to

I.C. $\qquad \xi = C - C_0 = \alpha \cos\left(\dfrac{\pi y}{L}\right)$

B.C. $\qquad \partial \xi / \partial y = 0, \qquad y = \pm L$

We obtain the solution we want by first substituting the general solution (eq. (3.49)) into initial conditions. This yields

$$\xi(y, 0) = \sum_{n=0}^{\infty} [A_n \sin(\lambda_n y) + B_n \cos(\lambda_n y)] = \alpha \cos\left(\frac{\pi y}{L}\right). \qquad (3.51)$$

This relationship is solved by setting all of the A_n terms and all but the first of the B_n terms to zero. We then set $B_0 = \alpha$, and $\lambda_0 = \pi/L$. It is straightforward to show that this choice of parameters also matches the boundary conditions since we only have a single cosine term left with a period $2L$ centred on the origin. Thus the concentration profile over time can be obtained by substitution back into eq. (3.49), as

$$C(y, t) = C_0 + \alpha \cos\left(\frac{\pi y}{L}\right) \cdot \exp\left(-\frac{\pi^2 D t}{L^2}\right). \qquad (3.52)$$

We see that the homogenization of a sinusoidal solute distribution simply involves an exponential decay with a time constant equal to $L^2/(\pi^2 D)$, as illustrated in Fig. 3.20.

Example 3.6: Alloy homogenization

Suppose that a Ni–Cr alloy casting is made with an average of 10 wt% Cr. It is found however that the local Cr composition varies between 7 and 13 wt% Cr (i.e. $10 \pm 3\,wt\%$) with a wavelength of about 20 µm. How long should the alloy be annealed at 1000°C in order to reduce the variation in composition to $\pm 1\%$?

The initial conditions can be described in terms of a cosine function, *i.e.*

I.C. $\qquad C = C_0 + \alpha \cos\left(\dfrac{\pi y}{L}\right) \quad$ or $\quad X^* = X_0^* + \alpha^* \cos\left(\dfrac{\pi y}{L}\right)$

where $X_0^* = 0.10$, $\alpha^* = 0.03$ and $L = 10^{-5}$ m. (Here, α^* is α converted to units of weight fraction.)

The diffusion coefficient for Cr in Ni at 1000°C is equal to

$$D = 1.1 \times 10^{-4} \exp\left(-\frac{272{,}000}{8.314 \times 1273}\right) = 7.59 \times 10^{-16}\,\mathrm{m^2/s}$$

The amplitude of the inhomogeneity at any time is given by $X^*(0, t) - X_0^*$. We therefore want to determine the time at which $X^*(0, t) = 0.11$. This involves a simple substitution into eq. (3.52) for $y = 0$:

$$0.11 = 0.1 + 0.03 \exp\left(-\frac{\pi^2 7.59 \times 10^{-16} t}{10^{-10}}\right)$$

Solving this for t, we find that the time needed is

$$t = -1.34 \times 10^4 \ln\left(\frac{0.11 - 0.1}{0.03}\right) = 1.47 \times 10^4\,\text{s.} = 4.1\,\text{h.}$$

3.6 Standard solutions to long-time transient diffusion problems

The solutions to a wide range of diffusion problems have been worked out over many years. There is therefore no need to keep re-inventing the wheel. Fortunately for us, these solutions have been made widely available in a number of handbooks.[¶] In this section we will look at a number of these solutions and think about how we can use them to solve a wide range of diffusion problems.

Our starting point is to remind ourselves of the necessary elements for the solution of boundary-value problems such as those involved with diffusion in solids. To set up a boundary-value problem we need to:

- define the governing equation,
- determine the appropriate geometry, and
- establish the initial and boundary conditions.

We shall examine a number of standard cases.

3.6.1 Uniform initial concentration C_i and fixed surface concentration C_s

Consider a plate of width $2L$ with boundary values given by

I.C.	$C = C_i,$	$-L < y < +L,$	$t = 0$
B.C.	$C = C_s,$	$y = \pm L.$	

The concentration at any position and time is given by

$$\frac{C - C_i}{C_s - C_i} = 1 - \frac{4}{\pi}\sum_{n=0}^{\infty}\frac{(-1)^n}{2n+1}\exp\left[-D\frac{(2n+1)^2\pi^2 t}{4L^2}\right]\cos\left[\frac{(2n+1)\pi y}{2L}\right].$$

$$(3.53)$$

This solution is obtained from Fick's Second Law using the separation of variables technique. If the same problem is solved using Laplace transform

---

[¶] The most important are those by H. S. Carslaw and J. C. Jaeger, *Conduction of Heat in Solids,* 2nd edition (Clarendon Press, Oxford, 1959) and a later book by J. Crank, *Mathematics of Diffusion,* 2nd edition (Clarendon Press, Oxford University Press, 1975)

methods the result is

$$\frac{C-C_{\mathrm{i}}}{C_{\mathrm{s}}-C_{\mathrm{i}}}=\sum_{n=0}^{\infty}(-1)^{n}\,\mathrm{erfc}\left[\frac{(2n+1)L-y}{2\sqrt{Dt}}\right]+\sum_{n=0}^{\infty}(-1)^{n}\,\mathrm{erfc}\left[\frac{(2n+1)L+y}{2\sqrt{Dt}}\right].$$

$$(3.54)$$

Although these expressions look completely different they do in fact give identical results. However, the first solution converges most rapidly at long times, close to equilibrium, while the second converges more rapidly at short times. Indeed, if the times are sufficiently short that the diffusion fields from each surface of the plate do not overlap, then we need take only the first term of eq. (3.54), (*i.e.* for $n=0$) which leaves two error-function solutions, one for each surface. At first sight these expressions look rather unwieldy, with a lot of variables to deal with. However, closer inspection quickly reveals that the parameters can be lumped together into only three dimensionless groups. The first of these is the normalized concentration $(C-C_{\mathrm{i}})/(C_{\mathrm{s}}-C_{\mathrm{i}})$ which appears on the left side of the equation. The second is the normalized time Dt/L^{2}. This appears in the exponential decay term of the first expression. The third group is the normalized position y/L. It is therefore possible to display these equations on a single plot, as shown in Fig. 3.21. Here the normalized concentration is plotted

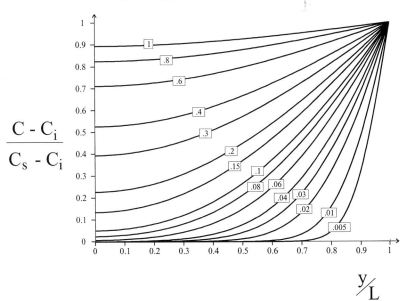

Figure 3.21 The normalized concentration $(C-C_{\mathrm{i}})/(C_{\mathrm{s}}-C_{\mathrm{i}})$ plotted as a function of the normalized distance y/L at various values of normalized time Dt/L^{2} for diffusion in a plane slab of thickness $2L$. The Dt/L^{2} values for each contour are given in the boxes. The initial solute concentration C_{i} is assumed to be uniform and the surface concentration is fixed at C_{s}.

against normalized position for a range of normalized times. The time contours on this plot cover the range over which the long-term transient solutions are generally valid. At shorter times, short-time solutions (which are simplified by using one boundary condition at infinity) can be used. At longer time we are sufficiently close to equilibrium. Within this time range we can use the plots directly without the need to use the mathematical expressions themselves.

Similar expressions are available for solid cylinders and spheres provided each has a uniform initial solute concentration C_i and a surface concentration fixed at C_s. (Since a sphere or a cylinder has only one surface, this condition gives only one boundary condition.) The second condition is obtained by recognizing that the flux at the centre of the body must be zero due to symmetry). The results for these geometries are similar.[¶] The normalized position for these geometries is given by r/a, the radial coordinate r divided by the radius a of the cylinder or sphere. The solutions are presented graphically in Figs. 3.22 and 3.23.

Example 3.7: Comparative diffusion rates into plates, cylinders and spheres

Consider a plate, a solid cylinder and a solid sphere with the same dimension (i.e., the radius of the sphere and cylinder equals the half-width of the plate).

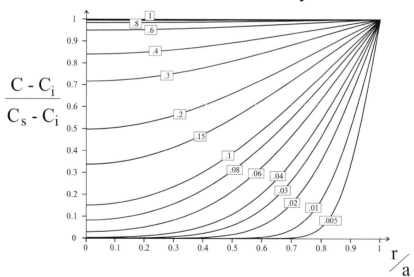

Figure 3.22 The normalized concentration $(C - C_i)/C_s - C_i)$ plotted as a function of the normalized distance r/a at various values of normalized time Dt/a^2 for diffusion in a solid cylinder of radius a. The initial solute concentration C is assumed to be uniform and the surface concentration is fixed at C_s.

¶ J. Crank, *Mathematics of Diffusion*, 2nd edition (Clarendon Press, Oxford, 1975), pp. 73 and 91

Each starts with the same initial solute concentration and is subjected to the same constant surface concentration. Compare the times taken for the concentration at the centre of each solid to increase half-way to the final value.

We need to find the time for the normalized concentration $(C - C_i)/(C_s - C_i)$ to become equal to 0.5 at a normalized position y/L or r/a equal to zero. This can be read directly from Figs. 3.21, 3.22 and 3.23 respectively. For the plate, the nearest time contour is 0.4, so that Dt/L^2 is about 0.38. For a cylinder, Dt/a^2 is about 0.21, while for a sphere, Dt/a^2 is about 0.13. While there is some error in interpolating between contours on these curves it is only about 1 or 2%, well within the uncertainty with which diffusion coefficients are known. We therefore conclude that diffusion is most rapid into a sphere and least rapid into a plate. The reason for this is clear if we think about these geometries. As solute diffuses towards the centre of the sphere the flux becomes concentrated. Therefore, less time is required.

Example 3.8: Hydrogen diffusion in steel wires

Suppose that steel wires with a 1 mm diameter (initially free of hydrogen) are exposed to a hydrogen atmosphere which fixes the surface concentration at 10^{-4} wt%. The temperature is 200°C. How long will it take until the concentration at the centre of the wire increases to 10^{-5} wt%? What will the concentration be at a position half-way from the centre to the surface, at this time?

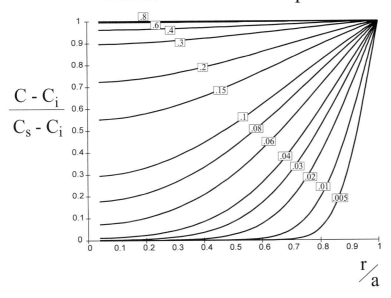

Diffusion in a Solid Sphere

Figure 3.23 The normalized concentration $(C - C_i)/(C_s - C_i)$ plotted as a function of the normalized distance r/a at various values of normalized time Dt/a^2 for diffusion in a solid sphere of radius a. The initial solute concentration C_i is assumed to be uniform and the surface concentration is fixed at C_s.

We first need to calculate the diffusion coefficient for H in Fe at this temperature. The required data is given in Appendix B. From this we find that (at 200°C)

$$D = 4.29 \times 10^{-10}\,\mathrm{m^2/s}.$$

We can write the initial and boundary conditions simply as:

I.C. $X^* = X_i^* = 0,$ $0 < r < a$

B.C. $X^* = X_s^* = 10^{-6}$ $r = a$

$$\frac{\partial X^*}{\partial r} = 0,$$ $r = 0.$

Here, $a = 0.5\,\mathrm{mm} = 5 \times 10^{-4}\,\mathrm{m}$. We want to know the time when $X^* = 10^{-7}$, i.e. when $(C - C_i)/(C_s - C_i) = (X^* - X_i^*)/(X_s^* - X_i^*) = 0.1$ at $r = 0$. From Fig. 3.22, this occurs when $Dt/a^2 \approx 0.085$. By substitution, the corresponding time is

$$t = 0.085\,\frac{a^2}{D} = 49.5\,\mathrm{s}.$$

We now want to know the corresponding concentration at $r/a = 0.5$. Once again we use Fig. 3.22, and follow the contour for $Dt/a^2 = 0.085$ out to $r/a = 0.5$. The value of $(C - C_i)/(C_s - C_i)$ is about 0.3. Thus, the concentration at this position is 3×10^{-7} or $3 \times 10^{-5}\,\mathrm{wt\%}$.

In addition to the specific concentration at a given location we often want to know the average concentration. The solution for this is obtained by simply integrating over the volume of the solid. The result for a plate is given by

$$\frac{\bar{C} - C_i}{C_s - C_i} = 1 - \frac{8}{\pi^2}\sum_{n=0}^{\infty}\frac{1}{(2n+1)^2}\exp\left[-D\,\frac{(2n+1)^2\pi^2 t}{4L^2}\right], \qquad (3.55)$$

or

$$\frac{\bar{C} - C_i}{C_s - C_i} = 2\sqrt{\frac{Dt}{L^2}}\left\{\frac{1}{\sqrt{\pi}} + 2\sum_{n=1}^{\infty}(-1)^n\,\mathrm{ierfc}\left(\frac{nL}{\sqrt{Dt}}\right)\right\}, \qquad (3.56)$$

where

$$\mathrm{ierfc}(z) = \int_z^{\infty}\mathrm{erfc}(\eta)\,\mathrm{d}\eta = \frac{1}{\sqrt{\pi}}\,\mathrm{e}^{-z^2} - z\,\mathrm{erfc}(z). \qquad (3.57)$$

The same dimensionless groups for concentration and time appear in these expressions as before. The average normalized composition is plotted as a function of normalized time in Fig. 3.24. This figure also contains plots for solid cylinders and spheres. This figure shows clearly that the rate of diffusion into a sphere is faster than that into a cylinder of the same radius and this is faster than diffusion into a plate of the same half-thickness.

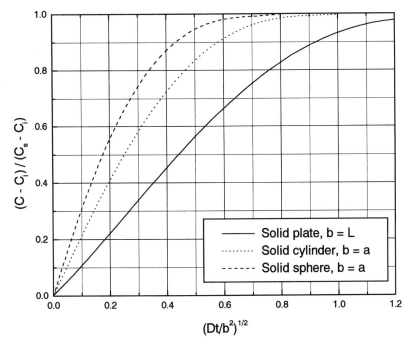

Figure 3.24 The normalized average concentration as a function of normalized time for three standard geometries. The dimension b is equal to the half-plate thickness L for a solid plate and equal to the radius a for a solid sphere or cylinder.

3.6.2 Uniform initial concentration C_i with fixed surface flux J^*

In some instances, the flux is fixed at the surface and not the concentration. The boundary values, for a plate of thickness $2L$, are then given by

I.C. $C = C_i,$ $-L < y < +L, \quad t = 0$

B.C. $J = -D\,\partial C/\partial y = J^*,$ $y = \pm L$

The solution to this problem is

$$C - C_i = \frac{J^*L}{D}\left\{ \frac{Dt}{L^2} + \frac{3y^2 - L^2}{6L^2} - \frac{2}{\pi^2}\sum_{n=1}^{\infty}\frac{(-1)^n}{n^2} \right.$$

$$\left. \times \exp\left[-D\left(\frac{n\pi}{L}\right)^2 t\right]\cos\left(\frac{n\pi y}{L}\right) \right\} \qquad (3.58)$$

This is plotted in Fig. 3.25. Note that the parameter $D(C - C_i)/J^*L$ is a dimensionless parameter. Once the diffusion profile extends to the centre of the plate, the fixed surface flux yields a series of parallel curves.

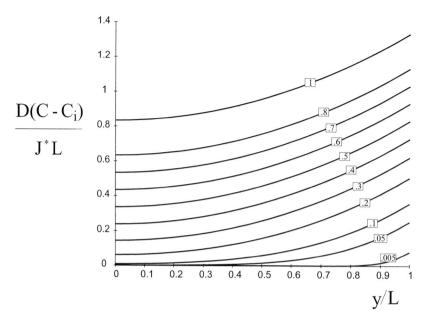

$$\frac{D(C - C_i)}{J^*L}$$

$$y/L$$

Figure 3.25 A normalized plot of concentration vs position for a fixed flux J^* at each surface of a plate. The boxes contain values of Dt/L^2.

In this case, the average concentration is easy to calculate. In general terms, the rate of increase with time is equal to the flux at the surface times the surface area divided by the volume of the body. Thus for a plate

$$\frac{d\bar{C}}{dt} = \frac{J_{y=L}}{2L} = \frac{J^*}{2L}.$$ (3.59)

In this case, since the surface flux is fixed, the average concentration is given simply by

$$\bar{C} = C_i + \frac{J^*}{2L} t$$ (3.60)

and increases linearly with time. Similar results are obtained for cylinders and spheres.

3.6.3   Uniform initial concentration but different boundary conditions

We now consider what happens if a plate of thickness L, with an initial uniform solute concentration, is subjected to boundary conditions on either side which fix the concentration at a constant value, but different on each side, *i.e.*

I.C.	$C = C_i$,	$0 < y < +L$,	$t = 0$
B.C.	$C = C_1$,	$y = 0$	
	$C = C_2$,	$y = L$.	

It is clear that in this case the system will move towards a steady state in which solute is continuously fed from the solute-rich side of the plate to the solute-poor side. In addition to the concentration profile and the average concentration, in this case it is also useful to calculate the amount of solute which has penetrated the plate per unit area as a function of time – call this Q. This is useful, for example, if the plate is a thin membrane separating two gases under pressure and we wish to know how much gas is leaking from one side to the other. The general solution is given by Crank.[¶] We will only consider the most common situation in which the initial concentration in the solid is zero, and the concentration on one side of the membrane remains at zero (*i.e.* $C_2 = 0$). In this case, the solution is given by

$$\frac{Q}{LC_1} = \frac{Dt}{L^2} - \frac{1}{6} - \frac{2}{\pi^2} \sum_{n=1}^{\infty} \frac{(-1)^n}{n^2} \exp\left[-D\left(\frac{\pi n}{L}\right)^2 t\right]. \qquad (3.61)$$

As time increases the exponential terms inside the summation decay and we are left with the first two terms. This establishes the steady-state condition. This result is plotted in Fig. 3.26. You can see that steady state is established after a normalized time Dt/L^2 of about 0.4.

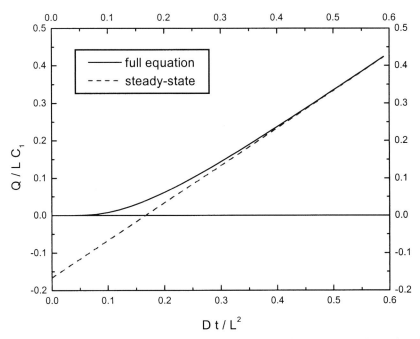

Figure 3.26 The amount of solute Q penetrating a membrane is plotted as a function of normalized time, for the special case of a membrane which is initially solute-free, and with concentration on one side fixed at zero.

¶ J. Crank, *The Mathematics of Diffusion*, 2nd edition (Clarendon Press, Oxford, UK, 1975), p. 51

3.6.4 Diffusion through a hollow cylinder or sphere

The last type of solution we will consider pertains to hollow cylinders and spheres. We will once again assume a uniform initial concentration of the solute. As in the previous example, we will assume that the two surfaces (*i.e.* the inner surface of radius a, and the outer surface of radius b) have fixed concentrations of different value. Thus

$$
\begin{array}{llll}
\text{I.C.} & C = C_i, & a < r < b, & t = 0 \\
\text{B.C.} & C = C_1, & r = a & \\
& C = C_2, & r = b &
\end{array}
$$

Solutions for the average concentration are shown in Figs. 3.27 and 3.28 (for cylinders and spheres respectively) for the special case in which $C_1 = C_2 = C_s$. It is clear that the difference between a hollow cylinder and a solid cylinder (or sphere) is not large, when $Dt/(b-a)^2$ is used as a normalizing parameter. This

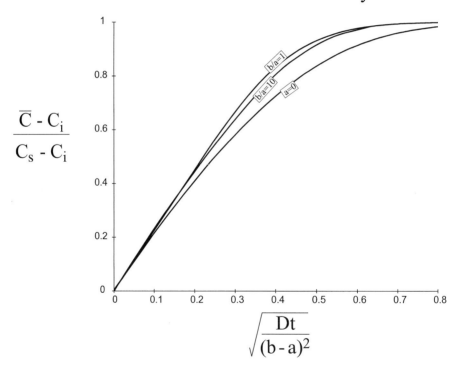

Figure 3.27 The average concentration in a hollow cylinder for a uniform initial concentration C_i and the surface concentration fixed at C_s on both inner and outer surfaces. The boxes contain values of b/a. (Note: b/a approaching 1 represents a thin-walled cylinder, which can be treated as a solid plate, while $a = 0$ represents a solid cylinder.)

suggests that most of the effect of the inner boundary is contained in the factor $(b - a)$. Beyond this, the normalized average concentration differs by at most 10%. A value of b/a approaching unity (corresponding to a solid plate), yields a solution which is negligibly different from a relatively thick-walled cylinder with b/a equal to 10. It would therefore appear that solutions of adequate accuracy can be obtained for hollow cylinders and spheres by treating them as linear plates of thickness $(b - a)$.

3.6.5 Other solutions

These are by no means the only diffusion problems for which solutions are available. While it is not possible to outline all of them here, the interested reader can find a wide selection in the books by Carslaw and Jaeger, and Crank. For example, one can find solutions related to homogenization involving arbitrary initial conditions and either fixed or periodic boundary conditions.

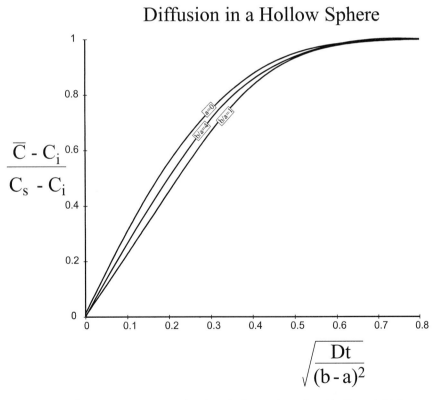

Figure 3.28 The average concentration in a hollow sphere for a uniform initial concentration C_i and the surface concentration fixed at C_s on both inner and outer surfaces. The boxes contain values of b/a. (Note: $b/a = 1$ represents a solid plate, while $a = 0$ represents a solid sphere.)

Solutions relevant to evaporation at a surface are also available. Finally, diffusion from a solution of limited volume is widely discussed. This is important if the source of the solute is depleted during a process, resulting in a continuously changing boundary condition.

3.7   **Further reading**

J. Crank, *The Mathematics of Diffusion*, 2nd edition (Clarendon Press, Oxford, UK, 1975)

P. Shewmon, *Diffusion in Solids*, 2nd edition (TMS-AIME, Warrendale, PA, 1989)

J. S. Kirkaldy and D. J. Young, *Diffusion in the Condensed State* (Institute of Metals, London, 1987)

R. J. Borg and G. J. Dienes, *An Introduction to Solid State Diffusion* (Academic Press, San Diego, CA, 1988)

3.8   **Problems to chapter 3**

3.1   (a) Beginning with Fick's First Law, set up a simple one-dimensional model and use it to derive Fick's Second Law.

   (b) Describe briefly the types of problems in which it is more appropriate to use Fick's First Law.

3.2   (a) The diffusion of matter into a solid from a plane surface at which the concentration is fixed at $C = C_s$, is described by the equation

$$C(y,t) = C_s \operatorname{erfc}\left(\frac{y}{2\sqrt{Dt}}\right)$$

Use this to show, as precisely as possible, the origin of the 'nominal' diffusion distance.

   (b) A solid is heated slowly from $0\,°C$ to $1000\,°C$ at the rate of $100\,°C/$hour. It is then held at $1000\,°C$ for 1 hour. Using the concept of a nominal diffusion distance, estimate the diffusion depth during (i) the heating period, and (ii) the hold period at $1000°C$.
(Assume $D_0 = 10^{-4}\,\mathrm{m^2/s}$, and $Q = 100\,\mathrm{kJ/mol}$, where $D = D_0 e^{-Q/RT}$, $R = 8.314\,\mathrm{J/mol\,K}$.)

3.3   Explain the significance of the parameter $(Dt)^{1/2}$, and give one example of its use in solving problems.

3.4   A thin layer of Cu is plated onto the surface of a Ni sheet. The sample is then annealed[¶] at $800\,°C$. Describe what measurements you would have to make and how you would plot the data in order to determine the diffusion coefficient for Cu in Ni. Give as precise a description as

---

[¶] For a definition of the term 'anneal' as used in this book, please see the footnote on p. 104.

possible, *i.e.* write down if possible, the equation on which the analysis is based.

3.5   A certain metal, call it element A, has a diffusion coefficient of $10^{-12}\,m^2/s$ at 1200°C and $10^{-10}\,m^2/s$ at 1300 °C.

(a) What is the activation energy for diffusion in metal A? Be sure to write down the units as well as the value.

(b) A thin layer of a radioactive tracer A* is plated onto the surface of the metal. It is then annealed at 1250 °C. Using the diffusion data given above, calculate how long it will take until half of the radioactive material has moved more than 100 μm from the surface.

3.6   You wish to measure the diffusion of oxygen in a silicate glass. This can be done using ^{18}O as a radioactive tracer, mixed with normal ^{16}O. The experiment is as follows:

(i)   The silicate is placed in a tube, which is exposed at one end. The tube is placed in a furnace.

(ii) Oxygen gas containing some ^{18}O is let in at the exposed end, at $t = 0$, up to a pressure of p, and the supply valve is closed.

(iii) The tube is left in the furnace at a fixed temperature T, for time t, to allow diffusion to occur. The tube is then quenched.

(iv) Thin slices are removed from the silicate bar. The weight and position of each is accurately measured.

(v) The number of radiation counts per second, N_c, is then measured for each slice.

(a) How would you analyze the data from this experiment to determine the diffusion coefficient for oxygen in the silicate? How would you plot the results to give a straight line from which D is determined? What assumptions are implicit in the analysis?

(b) If the volume of the gas chamber on the inlet side is quite small, then the curve is not a straight line. What shape does it have? Indicate how you would attempt to correlate for this effect?

3.7   You wish to measure the diffusion coefficient of a material using the radioactive tracer method. You are able to deposit $100\,mol/m^2$ of a radioactive tracer on the surface. You estimate that at a suitable temperature the diffusion coefficient is in the range of $10^{-10}\,m^2/s$. After annealing the material at this temperature you will be able to cut the piece into slices 0.1 mm thick. These will then be tested for radioactivity using a Geiger counter. The minimum concentration of tracer you can measure using this technique is $1000\,mol/m^3$. You want to ensure that you have at least 10 slices from which valid concentrations of the tracer can be measured. How long must you anneal the specimen?

3.8   Two copper rods are plated with a thin layer of radioactive copper and then butt welded together to form a metallurgical bond. The whole assembly is then annealed at 950 °C for 10 hours. Two slices are taken at 0.3 and 0.6 mm from the position of the radioactive layer. The ratio in the level of radioactivity between these two positions is 150. What is the diffusion coefficient of copper at this temperature?

3.9   (a) The solution for diffusion from a plane initial source containing an amount β of solute is

$$C(y, t) = \frac{\beta}{2\sqrt{\pi D t}} \exp\left(\frac{y^2}{4Dt}\right)$$

Use this to derive an expression for the concentration profile after a bar with solute at concentration C_0 is joined to a solute-free bar. Your solution should be applicable at relatively short times.

(b) Suppose $X_0^* = 15\,\mathrm{wt}\%$ and $D = 5 \times 10^{-11}\,\mathrm{m^2/s}$. Find the distance from the initial interface at which a concentration of $5\,\mathrm{wt}\%$ will be reached after annealing for 20 hours.

3.10   A thick plate of 0.3 wt% C steel is to be carburized at 930 °C. The carburizing gas used holds the surface concentration at 1 wt% C.

(a) Determine how long it will take for the C concentration to reach 0.6 wt% at a depth of 0.3 mm.

(b) How long will it take to double the depth at which this concentration is reached? (**Do not** re-calculate the answer from scratch.)

(c) By how much would the time in part (a) be changed if the temperature were raised by 50 °C? (As before, **do not** re-calculate the answer from scratch.)

3.11   A 2 mm thick silicon wafer is to be doped with antimony (Sb) in order to create a p-type region. You do this by passing a $SbCl_3/H_2$ gas mixture over the surface of the wafer at 1200 °C, which fixes the surface Sb concentration at $10^{23}\,\mathrm{/m^3}$. Suppose that you want the donor density (which is just the Sb concentration) to be greater than or equal to $3 \times 10^{22}/\mathrm{m^3}$, to a depth of 1 µm below the surface. Determine how long the wafer should be exposed to this atmosphere in order to achieve this. The diffusion coefficient for Sb in Si is given by $D = D_0 \exp(-Q/RT)$, where $D_0 = 1.3 \times 10^{-3}\,\mathrm{m^2/s}$ and $Q = 383\,\mathrm{kJ/mol}$.

3.12   Suppose that a fine metal wire of 1 mm diameter contains 0.005 wt% hydrogen. You wish to remove as much of this hydrogen as possible from the material. The wire is annealed in an atmosphere which establishes the equilibrium hydrogen level in the metal at 0.0008 wt%.

(a) If the diffusion coefficient for hydrogen in the metal is $2.5 \times 10^{-10}\,\mathrm{m^2/s}$, how long will it take for the average hydrogen concentration to drop to 0.001 wt%?

(b) How much longer will it take to lower the hydrogen concentration to 0.00085 wt%?

3.13 The change in the hydrogen potential of an aqueous solution can be measured by the change in the electrical resistance of a metal foil immersed in the solution. This is because the electrical resistance of the metal is proportional to the concentration of hydrogen. Suppose you wish to measure a sudden change in the hydrogen potential with this method. You wish to know the change in potential within a given time period, and to within a fixed accuracy.

(a) Set up a model which results in an expression (or a method) for obtaining the maximum width of the foil. Use the modelling methodology developed in Appendix D for this and outline each step in the process.

(b) If you wish to know the change in potential within 1 s, at an accuracy of better than 5%, then what is the maximum thickness of the foil? Use a diffusion coefficient of $10^{-9}\,\mathrm{m^2/s}$.

3.14 You wish to carburize 0.2 wt% C steel ball bearings. The balls have a diameter of 2.5 mm. You anneal them at 930 °C in an atmosphere which fixes the surface carbon concentration at 1 wt% C.

(a) How long will it take to raise the average C concentration to 0.6 wt%?

(b) What is the concentration at the centre of each ball at this time?

3.15 Suppose you have just learned to make wires of the high-temperature ceramic superconductor, $YBa_2Cu_3O_{7-x}$. You wish to optimize its superconducting properties by annealing the starting powder in oxygen at 900 °C prior to pressing it into wires. Suppose the powder particles average 10 μm in diameter. Calculate how long you would need to anneal it if you wish to ensure oxygen saturation, by determining the time to proceed 99% of the way towards equilibrium.

Note: A recent estimate of the diffusion coefficient of O in $YBa_2Cu_3O_{7-x}$ gives $D_0 = 7.3 \times 10^{-7}\,\mathrm{m^2/s}$ and $Q = 164\,\mathrm{kJ/mol}$.

3.16 Several steel companies produce grain-oriented steel for transformers, for example an Fe–3.2 wt% Si alloy. After the sheet has been cold-rolled to its final thickness of 0.257 mm, it is given an anneal to remove most of the carbon. Suppose that this anneal is done at 400 °C, and that the steel has an initial carbon content of 0.015 wt%. Further, suppose that this anneal is given in a CO/CO_2 atmosphere which fixes the surface concentration at 0.002 wt%.

(a) Determine how long it will take to reduce the average C content to 0.005 wt%. What will the C content be at the centre of the sheet at this time?

(b) Now suppose that because of faulty flow meters in the gas feed line, the pressure of CO_2 is only one-half what it should have been in part (a). Assuming that the CO pressure is correct, determine the average C concentration after the same time determined in part (a).

3.17   A steel is to be nitrided at 950 °C using an NH_3/H_2 mixture.

(a) Give an expression which tells you what the surface concentration of dissolved nitrogen will be. Be sure to define all the terms precisely.

(b) Suppose that a 1 mm thick plate is to be nitrided, and that the surface nitrogen level is fixed at 0.1 wt%. How long will it take until the average nitrogen content in the steel rises to 0.01 wt%. What will the concentration be at the centre of the plate at this time?

(c) Suppose that after some time the H_2 pressure accidentally doubles. Explain what will happen to the nitrogen concentration in the steel. (Be as precise as possible.)

3.18   It is important that copper pipes used in plumbing applications be kept free of hydrogen during brazing operations. Consider copper pipes with a 15 mm outer diameter and a 2 mm wall thickness. If brazing results in a temperature of 800 °C in the pipes, what is the maximum time this can be maintained if the average hydrogen level in the pipes is to be less than 20% of the saturation value?

3.19   Based on the data given in Problem 2.3, determine how long it will take to establish steady-state conditions. To do this find the time required until the rate of evaporation from the outside wall of the vessel is within 5% of the steady-state value. What is the pressure inside the vessel at this time? Based on your analysis, comment on the validity of the pseudo-steady-state assumption.

3.20   Porous nickel is used in a range of filter applications. This can be made by casting sheets of nickel spheres embedded in a polymeric binder. When the binder is burnt off it leaves a carbon residue on the surface of each particle. In order to give the material some strength the particles are heated to 600 °C for half-an-hour. During this time carbon can diffuse into the nickel. Phase diagrams show that the solubility of carbon in nickel at 600 °C is 0.2 atomic %. Determine the average carbon concentration in the nickel following this process. What will the carbon concentration be at the centre of each particle?

3.21   (a) Suppose that a Si wafer has been ion bombarded so as to greatly increase the diffusion over a layer extending 2 μm below the surface. If the wafer is placed in a PCl_3-containing atmosphere which fixes the surface concentration of P at 0.1 wt%, how long will it take until the average P concentration in the ion-bombarded layer is 0.08 wt%. (Assume that $D_{\text{P in Si}} = 10^{-16}\,\text{m}^2/\text{s}$.)

(b) If the diffusion coefficient outside the bombarded layer is $10^{-17}\,\text{m}^2/\text{s}$, estimate how much P will have diffused out of the layer in this time. You will need to make some simplifying assumptions in order to get a rough estimate. Make sure you state your assumptions.

3.22   During solidification, solute segregation occurs such that the last regions to solidify contain much higher levels of solute than the bulk of the material. Using a simplified model we might assume that the solute is enriched to a concentration C^* over a thin region of thickness δ and that these regions are spaced a distance $2L$ apart. We can assume for simplicity that outside of the enriched regions the initial solute concentration is equal to zero. The material can be homogenized by heat treatment at high temperature.

(a) Write down the initial and boundary conditions for this homogenization process.

(b) Let $\delta = 1\,\mu\text{m}$ and (using the first term of the series only) calculate the time required for the maximum solute concentration to drop to $\frac{1}{2}C^*$.

(c) At this time, what is the minimum concentration, if $L = 5\,\mu\text{m}$? (If you failed to get an answer for part (b) use a time of 1 hour).

(Assume a diffusion coefficient of the solute equal to $10^{-16}\,\text{m}^2/\text{s}$.)

3.23   A bottle of pungent perfume is opened about 10 m away from you and you notice the smell in a few seconds. Could the scent have diffused through the air to your nose or was it transported by convection? (D in air $\sim 10^{-4}\,\text{m}^2/\text{s}$.) Explain your reasoning.

Chapter 4

Applications to problems in materials engineering

'Cheshire Puss', she began, rather timidly . . .
'Would you tell me please, which way I ought to go from here?'
'That depends a good deal on where you want to get to,' said the cat.
'I don't much care where –' said Alice.
'Then it doesn't much matter which way you go,' said the cat.
Lewis Carroll, *Alice Through the Looking Glass*

We now turn our attention to the application of Fick's Laws for the solution of problems of interest to materials engineers. In this brief chapter the groundwork is laid through a discussion of the kinds of boundary conditions that can be applied.

4.1   Introduction

In the next few chapters we will build on our understanding of solutions to Fick's Laws, by learning how to apply them to a large variety of problems involving diffusion. This process involves a number of well-defined steps. These steps are required in order to turn a general statement of a problem into a framework which can be solved mathematically. The procedure we will adopt is generally applicable to all problems involving the solution of partial differential equations such as Fick's Laws. As we discussed in the last chapter, these equations fall under the general category of what are called 'boundary-value problems'. That is, we need to solve a differential equation (such as Fick's Second Law), subject to certain bounds. Because our equation only tells us how fast the concentration is changing, we need to give it a starting point, *i.e.* we need to define the concentration profile in the body at some starting time. This is called the *initial condition*. We also need to define what happens at various interfaces or surfaces. These are called *boundary conditions*. So we see that in order to properly and completely set up a boundary-value problem we need to:

- define the governing equation,
- determine the appropriate geometry, and
- establish the initial and boundary conditions.

4.2   Boundary conditions

Fick's Second Law is a *second*-order partial differential equation, *i.e.* it contains the second derivative of the concentration with respect to position. This means that we need *two* boundary conditions. What exactly is a boundary condition? Well, it describes the situation at some fixed location, valid for *all* time (or until the conditions change and we have to set up a new problem). For diffusion, a boundary condition involves either the concentration or some function of it. The most common boundary conditions either invoke a fixed concentration or a fixed flux (*i.e.* a fixed value of $\partial C/\partial y$). Determining the appropriate boundary conditions is by far the most interesting (and the most challenging) part of setting up diffusional boundary-value problems. We will therefore devote considerable effort to this subject. However, we will soon find that most boundary conditions fall into a small number of categories.

If the concentration is fixed at some location in the body this is generally an indication of thermodynamic equilibrium. For example, we may have a gas mixture of fixed partial pressures in contact with a solid surface. This will fix the solute concentration in the solid at the surface. Thus a CO/CO_2 gas mixture will control the C concentration at the surface of a steel specimen exposed to this gas. Or a PCl_3/H_2 gas mixture will control the P concentration at the surface of a silicon wafer over which it passes. In setting up such problems we generally assume that the surface is in a state of *local equilibrium*. This simply means that the interchange of atoms between the gas and the solid is quick enough that for the first few atomic layers at least the solid will attain the equilibrium value instantaneously, *i.e.* equilibrium is established locally at the surface. The same thing can occur internally at interfaces within a solid. If the interface separates two different phases (*e.g.* aluminum and θ-precipitates, or MgO and spinel) then equilibrium can be established at the interface. This is a local process since it only requires the transfer of a few atoms across the interface.

The *condition of local equilibrium* is a very powerful tool in setting up and solving diffusion problems. However, we must always keep in mind that it is an assumption, and as such it may not always be valid. We will discuss this further at a later stage (see, for example, Section 5.5).

The other common boundary condition is that of a fixed flux. This can arise in a number of ways. The first and most common is through symmetry, whereby the flux at a given point is the same in all directions. At any point of symmetry $\partial C/\partial y$ or $\partial C/\partial r$ is equal to zero, and thus the flux is zero. For example, the centre of a spherical ball-bearing into which carbon is diffusing uniformly from all directions is a point of symmetry. Symmetry is observed for cylindrical and planar geometries also. The fixed-flux boundary condition also arises if a surface or interface is impermeable. If the solute cannot penetrate the interface then the flux must be zero at the interface. Now this raises an interesting question. While the flux must be zero on one side of an impermeable interface does it have to go smoothly to zero? To put this both more precisely and more generally, must the flux be continuous across an interface? The easiest way of thinking about this is to ask what would happen if the flux were not continuous. For example, if a finite flux of solute arrives at a surface but none escapes then a build-up of the solute must result. In order to accommodate this excess solute, one of two things must happen. Either some new phase must form, or else the interface must move. For example, in the case of oxidation of a metal, when oxygen diffuses across the oxide layer to the metal/oxide interface it reaches what is essentially an impermeable interface (*i.e.* metals typically have rather low solubilities for oxygen). In this case, the oxygen reacts with the metal to form additional metal oxide and the interface moves (*i.e.* the thickness of the oxide scale increases). In other cases however, there may be no reaction available at an impermeable interface. In this case then the flux is restricted at the interface and must go smoothly to zero. So we can say that in the absence of a moving interface or the development of new phases, the diffusive flux is continuous across an interface. Thus, an impermeable surface requires that the flux goes smoothly to zero, and the boundary condition at an impermeable surface is one of zero flux.

The key thing to note about boundary conditions is that they embody most of the materials science aspects of a diffusion problem. This makes determining the correct boundary conditions for any given case one of the most challenging (and rewarding) parts of setting up and solving mass transport problems. Therefore, we will devote most of the remainder of this book to a discussion of specific problems and the boundary conditions that arise in each case.

Chapter 5

Applications involving gas–solid reactions

Though this be madness, yet there is method in't.
William Shakespeare, *Hamlet*

In this chapter we consider problems involving diffusion from a gas phase into the solid. The boundary conditions at the gas/solid interface are determined, assuming that local equilibrium exists, from a knowledge of thermodynamics. We will also consider an example in which the rate of reaction at the interface is the controlling step.

5.1 Introduction

We start by considering the interaction between gases and solid surfaces. This is a useful starting point for several reasons. First, this area includes a number of important technological problems in materials engineering. Second, these problems generally involve only a single diffusing species. This greatly simplifies matters. Examples of gas species which diffuse readily into solids include carbon and nitrogen in steel, hydrogen in many metals and ceramics, and water (H^+ or H_3O^+ ions) into glass. One common feature of these gases is that they involve rather small ions, which diffuse interstitially into crystalline solids. They therefore exhibit fairly high diffusion coefficients.

There is a related set of problems, involving diffusion from liquids into solids, which can be handled in the same way. For example, glasses can be strengthened by allowing K^+ or Ag^+ to diffuse into the surface, through ion exchange from a liquid. Similarly, ions can be leached out of a glass when in contact with a liquid.

In treating all of these situations, it is usually reasonable to assume that diffusion is much more rapid in the fluid (gas or liquid) than in the solid. As a result the fluid will not sustain a concentration gradient, *i.e.* the composition in the fluid at the interface will be the same as that in the bulk. In the case of gases, on which we will focus here, this means that the partial pressures of the gas at the surface of the solid is the same as that in the bulk gas. Thus it can be used to determine the boundary conditions for the problem.

5.2   **Simple gas–solid reactions**

Consider what happens when nitrogen gas is exposed to a solid. Some of the nitrogen will be dissolved in the solid, generally in atomic form. The reaction is therefore

$$(N_2)_g \rightleftharpoons 2\underline{N}.$$

The underline is used to denote a dissolved species. The equilibrium constant for this reaction can be written

$$K = \frac{(X_{\underline{N}}^o)^2}{p_{N_2}} = \exp\left(-\frac{\Delta G^o}{RT}\right), \tag{5.1}$$

where $X_{\underline{N}}^o$ is the equilibrium concentration (in units of mole or weight fraction) of dissolved nitrogen, p_{N_2} is the nitrogen gas pressure, ΔG^o is the standard free energy for the dissolution reaction, and R is the universal gas constant. Rearranging the equation, we get

$$X_{\underline{N}}^o = \sqrt{K \cdot p_{N_2}} = \sqrt{p_{N_2}} \exp\left(\frac{-\Delta G^o}{2RT}\right) \tag{5.2a}$$

or

$$C_{\underline{N}}^o = K_{N_2}^* \sqrt{p_{N_2}}. \tag{5.2b}$$

Note that while K is the equilibrium constant for this reaction, $K_{N_2}^*$ is the solubility of N_2 in the solid of interest.[¶]

These relationships (eqs. (5.2a) and (5.2b)) represent different forms of Sievert's Law and they are generally applicable to the dissociation of diatomic gases which diffuse into metals in atomic form. As you know, ΔG^o can be

---

[¶] We can define K^* either as the solubility at 1 atm, in which case it has units mol/m^3, or as the solubility constant with units of $mol/m^3 \, atm^{1/n}$, where n depends on the nature of the dissociation process of the gas entering the solid. Thus for a diatomic gas (such as N_2) which enters the solid in atomic form, $n = 2$ and K^* has units of $mol/m^3 \, atm^{1/2}$.

written in terms of its two components (*i.e.* enthalpy and entropy) as

$$\Delta G^{\circ} = \Delta H^{\circ} - T\Delta S^{\circ}.$$

By substituting this into Sievert's Law, we can isolate the temperature dependence:

$$X_{\underline{N}}^{\circ} = \sqrt{p_{N_2}}\, \exp\left(+\frac{\Delta S^{\circ}}{2R}\right) \cdot \exp\left(-\frac{\Delta H^{\circ}}{2RT}\right), \tag{5.3}$$

i.e. the temperature dependence is solely due to the standard enthalpy of dissolution.

Now, to be more precise in what we have just done, we should really have used equilibrium activity $a_{\underline{N}}^{\circ}$ in place of equilibrium concentration $X_{\underline{N}}^{\circ}$ in eqs. (5.2a) and (5.3). However, in almost all cases of gas dissolved in solids, the equilibrium solubility is low. The activity is therefore governed by Henry's Law. This states that the activity coefficient is a constant, and thus activity is proportional to concentration,

$$a_{\underline{N}} = \gamma_{\underline{N}}^{\circ} X_{\underline{N}}, \tag{5.4}$$

where $\gamma_{\underline{N}}^{\circ}$ is a constant, known as the Henrian activity coefficient. We can refine this by noting that at equilibrium, $a_{\underline{N}} = a_{N_2} = p_{N_2}$. In the standard state, $p_{N_2} = 1$ atm. Therefore, $a_{\underline{N}} = 1$ when the dissolved nitrogen is in equilibrium with N_2 in its standard state. Let us call this concentration $X_{\underline{N}}^{\text{sat}}$. Thus,

$$a_{\underline{N}} = 1 \quad \text{when } X_{\underline{N}} = X_{\underline{N}}^{\text{sat}}. \tag{5.5}$$

By equating eqs. (5.4) and (5.5) it is clear that

$$\gamma_{\underline{N}}^{\circ} = \frac{1}{X_{\underline{N}}^{\text{sat}}} \tag{5.6a}$$

and

$$X_{\underline{N}} = X_{\underline{N}}^{\text{sat}} a_{\underline{N}}. \tag{5.6b}$$

The value of $X_{\underline{N}}^{\text{sat}}$ is the solubility (expressed as a mole fraction) for nitrogen in the metal in the presence of 1 atmosphere of nitrogen gas. As such, it provides an alternative to eq. (5.2b) which expresses the solubility in terms of absolute units. Solubility data is generally available either through compilations of thermodynamic data or on phase diagrams. For example, the Fe–H phase diagram in Appendix C shows the solubility of $\underline{H}$ in Fe for $p_{H_2} = 1$ atm.

As noted earlier, the temperature dependence of $X_{\underline{N}}^{\circ}$ depends on the enthalpy for dissolution ΔH°. This can be either positive or negative. Consider, for example, the case of $\underline{N}$ in Fe (see Fig. 5.1). For the α- and δ-phase regimes,

$\Delta H^{\circ} > 0$ and $X_{\underline{N}}^{\circ}$ increases with temperature. In γ-Fe, however, $\Delta H^{\circ} < 0$ and $X_{\underline{N}}^{\circ}$ decreases with temperature.[¶]

Alternatively, compilations of solubility data can be tabulated. This is generally expressed using the solubility constant K^*. A small selection of such data is given in Table 5.1.

Table 5.1 Selected solubility data for gases in solids.

$$\left[C^{\mathrm{sat}} = (C^{\mathrm{sat}})_0 \exp\left(-\frac{Q_{\mathrm{sat}}}{RT} \right) \right]$$

Gas	Solid	$(C^{\mathrm{sat}})_0$(mol/m^3 at 1 atm.)	Q_{sat}(kJ/mol)
H_2	α-Fe	370	+27.1
H_2	γ-Fe	131	+28.1
H_2	Cu	144	+37.7
H_2	Ni	249	+15.1
He	CGW-7740^a	5.3×10^{-4}	−6.4
Ne	CGW-7740	7.4×10^{-4}	−5.9
H_2	CGW-7740	2.4×10^{-3}	−4.5

^{a}CGW-7740 is a product of Corning Glass Works, and is a common form of Pyrex glass.
Source: Adapted from J. Shelby, *Handbook of Gas Diffusion in Solids and Metals* (ASM International, Materials Park, OH, 1996).

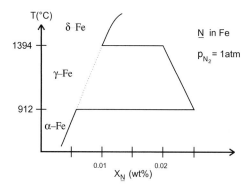

Figure 5.1 The solubility of nitrogen gas at 1 atm. pressure as a function of temperature in iron. Note the large degree of solubility in austenite and the negative temperature dependence. The high- and low-temperature phases of iron (δ-Fe and α-Fe respectively), which are both bcc, exhibit a continuous curve.

[¶] The reason why the solubility of nitrogen in δ-Fe appears to be an extrapolation of that for α-Fe is that both are really the same phase, *i.e.* they are both body-centred-cubic (bcc) iron.

Now that we understand something about gas–solid equilibria, we can use our knowledge to deal with problems involving diffusion. For example, nitrogen is often used to 'case' harden the surface of a steel component. We might want to know what depth of case is achieved by a given set of conditions. If we assume that local equilibrium exists between the gas and solid at the surface, the one boundary condition is given by Sievert's Law. Generally, the depth of a case is small compared to the thickness of the piece. If the part begins with a uniform nitrogen content, say X_0, then the conditions are given simply as:

I.C. $\qquad X_{\underline{N}} = X_0, \qquad\qquad y > 0, \quad t = 0$

B.C. $\qquad X_{\underline{N}} = X_{\underline{N}}^{\mathrm{o}}, \qquad\qquad y = 0$

$\qquad\qquad X_{\underline{N}} = X_0, \qquad\qquad y \to \infty.$

This is the familiar case of diffusion into a semi-infinite solid with a fixed surface concentration. The solution, as we have already seen (cf. eq. (3.29)), is

$$X_{\underline{N}} = X_{\underline{N}}^{\mathrm{o}} \, \mathrm{erfc}\left(\frac{y}{2\sqrt{Dt}} \right) \qquad\qquad (5.7)$$

Example 5.1: Solubility of hydrogen in iron

Using data from Table 5.1, determine the concentration of hydrogen dissolved in α-Fe at room temperature (20 °C), in equilibrium with a hydrogen gas partial pressure of 0.15 atm.

From the table we can determine the hydrogen concentration in the solid, in equilibrium with 1 atmosphere of hydrogen gas. It is

$$C^{\mathrm{sat}} = 370 \exp\left(-\frac{27{,}100}{8.314 \times 293} \right) = 5.45 \times 10^{-3} \, \mathrm{mol/m^3}$$

Since hydrogen gas is a diatomic molecule which dissociates to an atomic species in the solid, just like nitrogen, it also obeys Sievert's Law (eqs. (5.2) and (5.6)). Thus

$$C_{\underline{H}} = C_{\underline{H}}^{\mathrm{sat}} \sqrt{p_{H_2}} = 5.45 \times 10^{-3} \sqrt{0.15} = 2.11 \times 10^{-3} \, \mathrm{mol/m^3}.$$

In another situation, we might be more interested in diffusion of a gas, say hydrogen, into a thin solid film. In this case, the penetration is comparable to the film thickness $2L$. The boundary conditions are now given by local equilibrium at each surface:

B.C. $\qquad X_{\underline{H}} = X_{\underline{H}}^{\mathrm{o}}, \qquad\qquad y = \pm L.$

The solutions for this can be obtained by reference to those developed in Chapter 3. The composition profile is shown in Fig. 3.21, while the average composition of the film can be found using Fig. 3.24.

Note that we have used fractional concentrations rather than absolute concentrations throughout this treatment. Where necessary these will need to be converted.

5.3   Reactions involving gas mixtures

If we wish to control the surface concentration of a dissolved gas, then we must fix the partial pressure of the gas. This can be difficult using a single gas species. Greater control and flexibility can be achieved by using a gas mixture.

In the case of nitriding, for example, an NH_3/H_2 mixture is often used. The reaction of interest is

$$NH_{3(g)} \rightleftharpoons \underline{N} + \tfrac{3}{2} H_{2(g)} \tag{5.8}$$

and the equilibrium constant is

$$K = \frac{a_{\underline{N}} \cdot p_{H_2}^{3/2}}{p_{NH_3}} = \exp\left(-\frac{\Delta G^{\circ}}{RT}\right). \tag{5.9}$$

We now need to determine how $a_{\underline{N}}$ is related to the nitrogen concentration X_N.

For a mixed gas, such as NH_3/H_2, we can apply the same methodology as we did for the simple gas case. When $p_{NH_3} = p_{H_2} = 1$ atm., the equilibrium solubility must again correspond to $a_{\underline{N}} = 1$. We again define $X_{\underline{N}}^{sat}$ as the nitrogen solubility in equilibrium with the gases at 1 atm. Therefore,

$$X_{\underline{N}}^{\circ} = K \frac{p_{NH_3}}{p_{H_2}^{3/2}} X_{\underline{N}}^{sat}. \tag{5.10}$$

From this, we obtain boundary conditions at surfaces in contact with mixed gases. Other important examples include CO/CO_2 or CO_2/O_2 mixtures for carburizing procedures.

One of the recurring problems with developing surface layers by diffusion is the large composition gradient that results. The error-function solution, eq. (5.7), as we have seen, exhibits the steepest gradient at the surface. Ideally, we would like a surface layer in which the composition is constant and equal to the optimum value over a defined depth, and then decreases rapidly. In the case hardening of steels by carbon (carburizing) for example, the optimum carbon concentration in the case is about 0.8 wt%, over a depth of a millimetre or so. One way to achieve this involves a two-stage process, as illustrated in Fig. 5.2. Initially, a gas mixture is used which fixes the surface concentration at a high value, around 1.3 wt%. This speeds up the diffusion process. In the second step, the carburizing gas is removed or the temperature is lowered, and annealing[¶] is then continued. The carbon near the surface is therefore removed and a more rounded concentration profile is achieved. By

---

[¶] A note on terminology: The word 'anneal' is sometimes used with a broad meaning, as a synonym for 'heat treatment'. It is also used more specifically to denote a heat treatment designed to relieve stress in a material. Throughout this text I have used the term in the first, more general, manner.

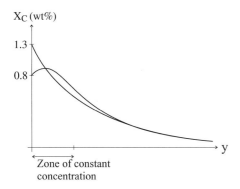

Figure 5.2 A two-step carburizing treatment produces a more uniform case. We first use a high-temperature anneal for which $(X_C)_s = 1.3\,\text{wt\%}$, followed by a lower-temperature anneal for which $(X_C)_s = 0.8 \quad \text{wt\%}$.

monitoring the time at temperature a profile close to optimum can be achieved.

5.4   Semiconductor doping

Semiconductor materials are doped with elements of a different valence in order to modify the number of conduction electrons or holes. In the early days of semicondutor device manufacturing, a common technique for accomplishing this involved passing a gas over the semiconductor at high temperature. Consider the most common semiconductor material, silicon. It can be doped with P to create a p-type semiconductor. To do this, PCl_3 gas is mixed with hydrogen (see Fig. 5.3) as a carrier gas. This mixture is then exposed to the surface of the Si wafer. The reaction is

$$PCl_{3(g)} + \tfrac{3}{2} H_{2(g)} \rightleftharpoons \underline{P} + 3\,HCl_{(g)}$$

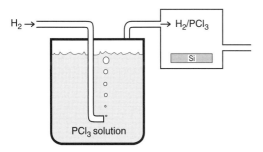

Figure 5.3 Silicon can be doped with phosphorus (to make a p-type semiconductor) by bubbling hydrogen through a solution with dissolved PCl_3 and passing it over a heated silicon substrate.

and the equilibrium constant is

$$K = \frac{a_{\underline{P}} \cdot p_{HCl}^3}{p_{PCl_3} \cdot p_{H_2}^{3/2}} \tag{5.11}$$

The p_{H_2}/p_{PCl_3} ratio can be varied by modifying the solution through which the H_2 gas is bubbled. This can be manipulated to obtain the desired level of $\underline{P}$ doping in the silicon.

One of the problems with this method of doping semiconductors is that a non-uniform distribution of P results. In semiconductor devices, better results could be achieved if the distribution were more uniform. One approach is to develop a two-stage heat treatment process, similar to that discussed in the previous section for the carburizing of steel. Another approach is to modify the diffusion coefficient over a limited depth beneath the surface of the wafer. This can be done using ion bombardment. As ions are injected with high energy into a crystal, they displace many of the atoms from their normal positions. This increases the density of defects in the structure, and thus the diffusion coefficient. For example, 100 keV protons produce damage in silicon to a depth of about 1 micron.

How do we model this? Well, let's assume that the diffusion coefficient is increased by a constant factor α over a given depth within the material. It then falls to the nominal value for the material below this depth (see Fig. 5.4).

Since there is no reaction at the interface between the two regions, the flux must be continuous across it, i.e.

$$J(y = L) = -\alpha D \left(\frac{\partial C}{\partial y}\right)_{y=L-} = -D \left(\frac{\partial C}{\partial y}\right)_{y=L+}. \tag{5.12}$$

Since $\alpha \gg 1$, the concentration gradient must be much smaller on the damaged side of the interface. If α is sufficiently large, we can treat the interface as if it

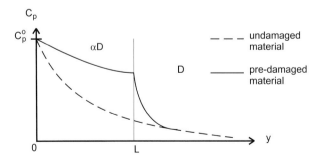

Figure 5.4 If the diffusion coefficient is increased by a factor α over a depth L due to ion bombardment the penetration of solute will be much more rapid into this region.

were impermeable. This produces a simple boundary condition:

$$\frac{\partial C}{\partial y} \approx 0 \quad \text{at } y = L.$$

This is combined with the condition of constant concentration fixed by equilibrium at the surface of the wafer. These two conditions are the same as those for a plate of thickness $2L$, with a fixed concentration on each surface. Thus the solutions are readily available.

Given the very fine scale on which semiconductor doping is performed in modern silicon devices, vapour deposition and diffusion processes are no longer used. They simply cannot offer the control of composition that is required.

5.5   Gas/solid interfaces without local equilibrium

While we have tended to assume that the condition of local equilibrium is always valid, this is not universally so. Occasionally the rate of dissolution of a gas into a solid is limited by the rate of absorption at the surface. Consider the case of hydrogen. There are three separate steps involved. The reactions for each step are:

$$\begin{array}{lll} H_{2(g)} & \rightleftharpoons H_{2(ads)} & (1) \quad K_1 \\ H_{2(ads)} & \rightleftharpoons 2H_{(ads)} & (2) \quad K_2 \\ 2H_{(ads)} & \rightleftharpoons 2\underline{H} & (3) \quad K_3. \end{array}$$

First, the diatomic hydrogen molecule is physically adsorbed onto the surface (reaction 1). It then dissociates into atomic hydrogen, again adsorbed on the surface (reaction 2). Finally, the hydrogen atoms are chemically absorbed into the surface (reaction 3). There is an equilibrium constant associated with each reaction, and for the overall reaction,

$$H_{2(g)} \rightleftharpoons 2\underline{H} \qquad K \tag{5.13}$$

which is $K = K_1 \cdot K_2 \cdot K_3$. Since the three processes involved are sequential (to which we can add the fourth sequential process of diffusion into the solid), the rate of the overall reaction is governed by the slowest process. Therefore when we assume that the surface is at local equilibrium what we really assume is that diffusion away from the surface is the slowest of these four processes.

In some materials, this is not always a valid assumption. For example, the diffusion of interstitial atoms can be very rapid. Thus when hydrogen diffuses into iron (or steel), reaction 2 (the dissociation of molecular hydrogen into atomic hydrogen on the surface) is rate-limiting at temperatures below about 200 °C. Let us consider the consequences of this. The flux across the interface will be limited by the rate of dissociation, which is proportional to the surface concentration of adsorbed hydrogen

$$J_s = k_d C_{H_2(ads)}$$

where k_d is the rate constant for this dissociation process. Since reaction 1 is rapid, the concentration of adsorbed H_2 molecules on the surface is in equilibrium with the bulk gas:

$$C_{H_2(ads)} = K_1 \, p_{H_2}.$$

This means that the flux will be a constant

$$J_s = k_d \, K_1 \, p_{H_2}. \tag{5.14}$$

The diffusion flux away from the interface is given by Fick's First Law. Therefore

$$\frac{\partial C_{\underline{H}}}{\partial y} = -\frac{k_d K_1}{D_H} \, p_{H_2} = -k^*, \tag{5.15}$$

where D_H is the diffusion coefficient for H in Fe. This equation now provides one of the boundary conditions for this problem. Let us consider a specific example. Suppose that a wire, radius a, is initially hydrogen-free. It is then placed in a furnace containing a fixed partial pressure of hydrogen at a temperature below 200 °C. The boundary-value problem can be stated as follows.

Geometry: cylindrically symmetrical

I.C. $C_{\underline{H}} = 0,$ $r \le a, \quad t = 0$

B.C. $\dfrac{\partial C_{\underline{H}}}{\partial r} = 0,$ $r = 0$

               $\dfrac{\partial C_{\underline{H}}}{\partial r} = -k^*,$ $r = a$

The solution for diffusion into a solid plate with a fixed surface flux is shown in Fig. 3.22. The solution for a cylinder is similar. It is shown schematically in Fig. 5.5. We see a set of concentration profiles that are parallel.

It should be clear, however, that these boundary conditions cannot be maintained until equilibrium is reached. As the solid wire fills up with hydrogen the concentration at the surface rises until eventually it reaches the equilibrium value. It cannot then continue to rise above this value. Instead, the kinetic limitation is removed and a condition of local equilibrium is established. Thus the second boundary condition above is replaced by $C_{\underline{H}} = C_{\underline{H}}^o$ at $r = a$. This transition is shown schematically in Fig. 5.5. This is one example of a problem in which circumstances change as a process progresses, such that the boundary conditions that apply when the process is begun are no longer relevant. New boundary conditions must then be formulated in response to

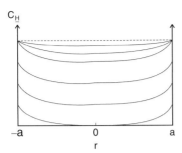

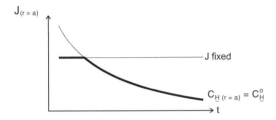

Figure 5.5 A set of diffusion profiles for the case in which the flux at the surface is held constant due to reduction kinetics. Once the surface concentration reaches the equilibrium value the surface flux starts to decrease.

the changed circumstances. It is important to realize that this is possible, and to anticipate it.

5.6   Further reading

J. E. Shelby, *Handbook of Gas Diffusion in Solids and Melts* (ASM International, Materials Park, OH, 1996)

5.7   Problems to chapter 5

5.1   A cylindrical vessel of radius R and length L is divided into two isolated sections by placing a metal plate of thickness δ at a distance d from the left end (see diagram overpage). You perform an experiment in which the right chamber of the vessel is evacuated, while the left one is filled with one atmosphere of hydrogen gas. The vessel is then isolated from the rest of the system (*i.e.* no gas can enter or leave). Assume the metal plate is originally hydrogen-free.

(a) Draw a diagram which illustrates the initial conditions. Be as quantitative as possible.

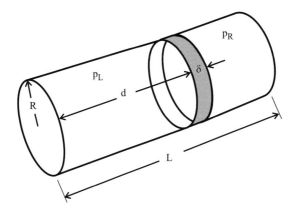

(b) Draw a diagram which illustrates the final conditions. Be as quantitative as possible.

(c) Draw a set of schematic curves for the concentration of hydrogen in the metal plate, for times ranging from $t = 0$ to $t = \infty$. Indicate any assumptions you make.

(d) Describe the boundary conditions for this process. Indicate any assumptions you make.

(e) Draw a figure indicating how the total amount of H_2 gas in the right-hand chamber will vary with time.

(f) Part-way through this process, a 'pseudo' steady state will develop. Derive an expression for the rate of pressure increase in the right-hand chamber during this period.

(g) If the diffusivity of hydrogen in the metal is not constant, but decreases with $\underline{H}$ concentration, how will this affect the concentration profile during the pseudo-steady-state period? If you could measure this profile, how could you use it to measure D as a function of $\underline{H}$ concentration?

5.2   SiC oxidizes slowly when exposed to air at high temperature. The process involves the growth of a silica film on the surface. The reaction at the SiC/SiO_2 interface is

$$SiC + \tfrac{3}{2}O_2 \rightarrow SiO_2 + CO.$$

For film growth to occur, O_2 and CO each must diffuse through the film. The rate of growth is controlled by the diffusion of O_2.

(a) Develop expressions to describe the O_2 concentration in the SiO_2 film at:
(i) the outside surface
(ii) the SiO_2/SiC interface.

Be as precise as possible, using assumptions where needed. Be sure to justify your assumptions.

(b) For the oxidation of Si_3N_4 the situation is more complicated, as shown below.

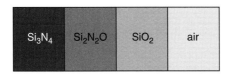

Assuming that film growth occurs under pseudo-steady-state conditions, what are the combination of parameters (Si_2N_2O and SiO_2 film thickness, diffusivities, etc.) which determine whether diffusion through the Si_2N_2O or the SiO_2 film controls the overall kinetics.

Note: You may assume that diffusion of O_2 is controlling in the Si_2N_2O film also.

5.3   A thin wire of iron, 1 mm in diameter, contains 0.005 wt% hydrogen. You wish to remove as much of the hydrogen as possible by annealing it an atmosphere containing a low hydrogen partial pressure of 0.01 atmospheres at room temperature. The solubility of atomic hydrogen in iron at this temperature in 1 atmosphere of hydrogen gas is 10^{-4} wt%

(a) What is the solubility of hydrogen in the annealing atmosphere?

(b) If the diffusion coefficient for H in the metal at this temperature is 2.5×10^{-10} m^2/s, how long will it take until the average hydrogen concentration drops to 10^{-4} wt%? What will the concentration be at the centre of the wire at this time?

(c) Repeat the above calculation for an average composition of 2×10^{-5} wt%.

5.4   A certain metal hydride MH_2 is being considered for hydrogen storage applications. It is proposed to store the hydride as a powder in copper-coated steel canisters. You wish to set up an experiment to measure the permeability of hydrogen through the canister. Suppose that the canister is a sphere 10 cm in diameter and that is placed in a vacuum furnace at 1000 °C. The hydride decomposes according to the reaction

$$MH_{2(s)} \rightleftharpoons M_{(s)} + H_{2(g)} \quad \Delta G_1^o(1000\,°C) = -7\,kJ/mol.$$

It then dissolves in the copper according to the reaction

$$H_{2(g)} \rightleftharpoons 2\underline{H} \qquad \Delta G_2^o(1000\,°C) = +50.7\,kJ/mol.$$

You measure the amount of hydrogen gas coming through the canister walls per unit time. Separate measurements have shown that the maximum rate of hydrogen evolution from the hydride powder is 4000 moles/m³ s.

You are concerned that if the copper coating is too thin you will not be able to supply hydrogen gas quickly enough to establish local equilibrium at the gas/copper interface. Estimate how thick the copper coating must be in order to ensure that this not a problem.

Assume that: (i) the metal M does not react with or dissolve in copper;

   (ii) the diffusion of atomic hydrogen in copper at 1000 °C is approximately 10^{-8} m²/s;

   (iii) all of the resistance to diffusion is provided by the copper (*i.e.* ignore the steel).

5.5   Consider the decarburization of a sheet of plain carbon steel, 5 mm thick, containing 0.32 wt% carbon. You are given a target composition of 0.01 wt% C (or less) everywhere in the final sheet, and two alternative treatments, both of which involve the batch exposure of the steel sheet to a decarburizing gas atmosphere which sets the surface carbon concentration at or near zero.

   (i) The first takes place at 800°C, a temperature that places the initial alloy composition precisely on the $(\alpha + \gamma)/\gamma$ equilibrium boundary.

   (ii) The second takes place at 940°C.

   (a) For each of these two processes, evaluate the time for decarburization of the steel sheet to the final required level. Your analysis should highlight the differences between the two processes and your answer should include several schematic composition vs distance plots for each process.

   (b) In the design of decarburization processes, it is desirable to adjust the gas-phase composition such that the carbon potential is close to zero, and also to leave the final surface bright and free of reaction products. You are given the choice of two gas mixtures: CO/CO_2; H_2/H_2O. Discuss, in as much detail as possible, the merits of each candidate mixture, and give the decarburizing reaction and the form of the gas pressure dependence in each case.

5.6   A spherical pressure vessel of 1 m diameter is composed of low-carbon steel, and is fabricated with a wall thickness of 1 mm. Against your best advice, it is being used to contain pure hydrogen gas at a pressure of 100 atm. and a temperature of 400 °C.

   (a) Estimate the time required to bring the internal pressure to 10 atmospheres. In your answer, be sure to state all of the assumptions made.

(b) State the time dependence of the hydrogen loss from the vessel, *i.e.* is the rate linear in time, parabolic, cubic, or some other power law?

(c) In an attempt to preserve hydrogen (and perhaps human lives), you elect to contain this spherical container in a copper cylinder of height 1.5 m and diameter 1.2 m. The copper vessel wall thickness is sufficient to effectively stop hydrogen loss from the outer container. First, state the final equilibrium pressures in the two (inner and outer) vessels. Second, estimate the time required to bring the inner vessel pressure to within 10% of its final value.

5.7   A steel ingot containing 1 wt% carbon was held in a soaking pit at 950 °C for 24 hours. Since the furnace gases were slightly oxidizing, some decarburization of the ingot occurred. Assume that the C concentration at the surface is zero. Plot a graph of the carbon distribution in the ingot after 24 hours exposure, and suggest what thickness of decarburized metal should be removed.

5.8   The solubility of hydrogen in solid copper at 1000 °C is 1.4 parts per million (ppm) (by mass) under 1 atm of hydrogen pressure.

(a) Determine the time for hydrogen to reach a concentration of 1.0 ppm at a depth of 0.1 mm in a large chunk of copper initially containing no hydrogen if the copper is subjected to 2 atm pressure of H_2 at 1000 °C.

(b) If a copper foil, 0.2 mm thick, is equilibrated with hydrogen at a pressure of 4 atm at 1000 °C, then placed in a perfect vacuum at 1000 °C and held for 1 minute, calculate the concentration of hydrogen at the centre of the foil.

5.9   Several thousand cylindrical gear shafts 0.3 m long ×25 mm diameter, made from 1020 steel are to be case-hardened in a reducing atmosphere containing not less than 98% CO at 925 °C. The case-hardened depth required is 1.5 mm (*i.e.* the carbon concentration at 1.5 mm below the surface should be 0.8 wt%). Calculate:

(a) the time required to achieve such a case-hardened depth;

(b) the total consumption of carbon monoxide during the period, due to depletion by the reaction $2CO \rightarrow \underline{C} + CO_2$;

(c) the mean CO flow rate (litres/hour at N.T.P. (0°C, 1 atm.)) that must be provided to the furnace to maintain the CO_2 content in the gases at no more than 2%.

Chapter 6

Heat treatment of binary alloys

There is something fascinating about science. One gets such wholesale returns of conjecture out of such a trifling investment of fact.

Mark Twain, Life on the Mississippi

This chapter introduces a wide range of examples related to the heat treatment of binary alloys. The process involved is very simple – an alloy of fixed overall composition is subjected to a given temperature–time cycle. However, the analysis can be quite complex. Our guide to the various possibilities is the appropriate binary phase diagram, which summarizes the equilibrium conditions for the system. We will almost always assume that local equilibrium is established at interfaces, the boundary conditions thus being given by the phase diagram. The problems of interest include the dissolution and the growth of precipitates and the growth of lamellar structures such as pearlite. We will briefly consider how the analysis method can be modified to treat systems involving a third component.

6.1   Introduction

We now wish to consider what happens when a multi-component material, initially at equilibrium, is subjected to a change of temperature. This is clearly related to the process of heat treatment of materials. Depending on the temperature change involved particles may dissolve, they may be precipitated or they may change in size and volume fraction. A rather beautiful example is

shown in Fig. 6.1, which shows TiN precipitates in a Mo–Ti–N alloy. At sufficiently high temperature a simple solid solution exists. Upon heat treatment at a
lower temperature, however, fine TiN precipitates, in the form of plates about
100 nm in diameter, develop. What actually happens depends on thermodynamics and kinetics. We are fortunate to have the thermodynamic information we need summarized neatly in the form of phase diagrams. These tell us
what phases will be present once the material reaches equilibrium. They also
tell us the proportions in which each phase will be present. Phase diagrams
also provide useful information in understanding the *kinetics* (or speed) with
which the system approaches a new equilibrium after a change in conditions.
This is because most interfaces between different phases are able to establish
local equilibrium at the boundary rather quickly. The phase diagram also tells
us what form this local equilibrium will take.

In what follows we will postulate a number of scenarios of interest in
understanding phase transformations in solids. We will then attempt to
describe the problem in detail. In particular, when describing the boundary
conditions, we will generally assume that local equilibrium exists at all interfaces between two different phases.

6.2   Changing temperature within a two-phase field

We start off with a simple case. Suppose that an alloy is at equilibrium within a
two-phase field, as illustrated by the phase diagram in Fig. 6.2.[¶]

Of the two phases involved, one (the α-phase) has an equilibrium composition which depends on temperature, while the other (the β-phase) does not
change its composition with temperature. This is not as unusual as it may
seem. Many intermetallic compounds have a narrow range of stoichiometry
which is temperature insensitive. Thus this scenario applies to important
materials such as plain carbon steels (the $Fe–Fe_3C$ phase diagram) and *age-
hardenable* aluminum alloys (*e.g.* the $Al–Al_2Cu$ phase diagram). We will
denote the overall solute content of the alloy by C_0. At the initial temperature
T_1, α-phase of composition $C_{\alpha 1}$ is in equilibrium with β-phase of composition
C_β. We now want to consider what happens if we lower the temperature to T_2.
As shown in Fig. 6.2, the equilibrium solute content of the α-phase decreases
with decreasing temperature and is equal to $C_{\alpha 2}$ when the temperature is T_2.
Since the composition of the β-phase is independent of temperature we need

---

[¶] Throughout this chapter and elsewhere we will make extensive use of binary phase
diagrams. It is therefore essential that you are able to read and interpret such diagrams
with ease. If you are not familiar with phase diagrams, or feel you need a refresher, I
suggest you consult *Engineering Materials 2* by M. F. Ashby and D. R. H. Jones
(Pergamon Press, Oxford, 1986). Appendix 1 – 'Teach yourself phase diagrams' provides
an excellent practical guide to this subject.

only worry about diffusion in the α-phase. We can start by asking qualitatively what will happen once the temperature is changed. Well, we have already seen how the composition of the α-phase will change – it decreases. This must occur by diffusion of solute towards the interface, as shown schematically in Fig. 6.3. This solute is absorbed at the interface by creating more β-phase. Thus the interface moves. Eventually, a new equilibrium will be reached in which all of the α-phase has a composition $C_{\alpha 2}$, but with less α and more β than when we started.

For the time being however, we will ignore the interface motion. We can now write down the initial, final and boundary conditions for this problem.

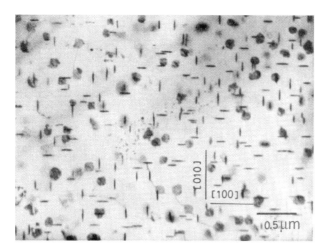

Figure 6.1 TiN preciptates in a Mo–Ti–N alloy. The precipitates form as fine platelets parallel to the [100] planes in the matrix (N. E. Ryan, W. A. Soffa and R. C. Crawford, Orientation and habit plane relationship for carbide and nitride precipitates in molybdenum. *Metallography*, **1**, 195 (1968)).

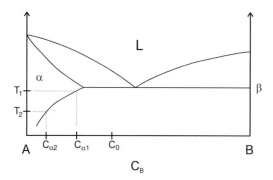

Figure 6.2 A schematic binary phase diagram for a simple eutectic system. Note that we have assumed that the β-phase exhibits a constant composition at all temperatures.

6.2.1   Solution valid at short time

At short time we can assume that the diffusion profile does not change close to the ends of the specimen. We therefore get one boundary condition from the assumption of local equilibrium at the α/β interface, while the other comes from assuming that the composition is unchanged far away from the interface. The conditions can be written in the following form:

$$I.C. \qquad C = C_{\alpha 1}, \qquad\qquad y < 0$$
$$B.C. \qquad C = C_{\alpha 2}, \qquad\qquad y = 0$$
$$\qquad\qquad C = C_{\alpha 1}, \qquad\qquad y \rightarrow -\infty.$$

Note that these conditions apply only to the α-phase $(-\infty < y < 0)$. No diffusion occurs in the β-phase, which we can regard as a sink for excess solute atoms. The solution appropriate to a *planar* geometry with these boundary conditions is the one with which we are by now familiar, based on the error function. It is

$$\frac{C(y,t) - C_{\alpha 1}}{C_{\alpha 2} - C_{\alpha 1}} = 1 + \mathrm{erf}\left(\frac{y}{2\sqrt{Dt}}\right). \tag{6.1}$$

You can easily convince yourselves, by substitution, that this solution matches the initial and boundary conditions and that it has a form which is consistent with the schematic diffusion profiles shown in Fig. 6.2.

6.2.2   Solution valid at long time

A more interesting problem involves thinking about a periodic arrangement of plates (*lamellae*), in which two phases are close enough for the diffusion fields to overlap. Such is the case with plates of ferrite and cementite in a pearlitic steel microstructure (shown in cross-section in Fig. 6.4(a)). This is shown schematically in Fig. 6.4(b) in which ferrite is denoted by α and cementite,

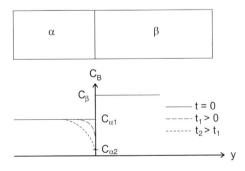

Figure 6.3 A schematic illustration of diffusion towards an interface for two semi-infinite bars joined together, following a temperature change.

Fe$_3$C, is denoted by β. The difference between this case and the previous one is only in the second boundary condition. We clearly do not have an infinite boundary. Instead, the conditions of local equilibrium define the boundary condition on each side of each α-phase plate. Thus,

I.C. $C = C_{\alpha 1},$ $-L < y < +L$

B.C. $C = C_{\alpha 2},$ $y = \pm L.$

Alternatively we note that, due to symmetry, the flux will always be zero at the centre of the plate. We can therefore describe the problem using the following set of conditions:

I.C. $C = C_{\alpha 1},$ $0 < y < L$

B.C. $C = C_{\alpha 2},$ $y = L$

$\partial C/\partial y = 0,$ $y = 0.$

It is important to realize that these two sets of conditions describe the same problem and lead to exactly the same result.

(a)

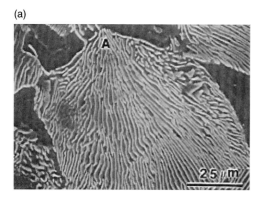

(b)

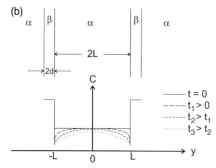

Figure 6.4 (a) The growth of pearlite in a 1060 steel (P. R. Howell, The pearlite reaction in steels: mechanisms and crystallography, *Mater. Charact.*, **40**, 225 (1998). (b) The diffusion profile for a series of alternating plates of the α- and β-phase, following a temperature change.

This is a standard problem of the filling up of a plate with a fixed surface concentration. The solution is given in Chapter 3 (see Fig. 3.21 and eqs. (3.53) and (3.54)).

Example 6.1: Thickening of cementite plates in pearlite

Consider the Fe–C (actually the Fe–Fe$_3$C) phase diagram (see Appendix C). At temperatures below 727 °C the two stable phases are α-Fe (ferrite) and Fe$_3$C (cementite). Suppose that a pearlitic steel with 0.77 wt% C has been equilibrated at 700 °C. The temperature is then lowered to 450 °C.

(a) What are the initial and boundary conditions for this problem?

These can be read directly from the phase diagram, assuming that local equilibrium conditions can be assumed at the interface between ferrite and cementite. Thus, the conditions are given by:

I.C. $X^* = X^*_{\alpha1} = 0.015\,\mathrm{wt\%}$, $0 < y < L$

B.C. $X^* = X^*_{\alpha2} = 0.001\,\mathrm{wt\%}$, $y = L$

$\partial X^*/\partial y = 0$, $y = 0$.

(b) Assuming that the interlamellar spacing $2(L+d)$ is equal to 10 μm, what is the geometry and length scale appropriate to this problem?

Clearly this is a one-dimensional problem involving planar symmetry. We can assume, as noted above, that the effect of interfacial motion is negligible. Therefore the problem is one of diffusion of carbon out of the ferrite plates of width L. We need to determine a value for L. From the Lever rule the weight fraction of ferrite in the pearlite is equal to $f^*_\alpha = (X^*_\beta - X^*_0)/(X^*_\beta - X^*_{\alpha1}) = (6.7\text{–}0.77)/(6.7\text{–}0.015) = 0.89$. This can be converted to volume fraction using the relationship (see Table 2.1, p. 30)

$$\varphi_\alpha = \frac{f^*_\alpha/\rho_\alpha}{f^*_\alpha/\rho_\alpha + (1 - f^*_\alpha)/\rho_\beta}$$

where ρ represents the density of each phase. In this case because the densities of the two phases are so similar ($\rho_\alpha = 7860\,\mathrm{kg/m^3}$, $\rho_\beta = 7690\,\mathrm{kg/m^3}$) this does not significantly alter the result, *i.e.* $\varphi_\alpha = L/(L+d) \approx 0.89$. Therefore we find that the width of the ferrite plates is equal to 8.9 μm.

(c) How long will it take until the carbon concentration in the centre of the ferrite plate has fallen to $2X^*_{\alpha2}$? This might be used to approximate the time required to approach the new equilibrium condition.

We need to determine the time at which $X^*(0, t) = 2X^*_{\alpha2}$. In terms of normalized variables this represents $[X^*(0, t) - X^*_{\alpha1}]/(X^*_{\alpha2} - X^*_{\alpha1}) = [C(0, t) - C_{\alpha1}]/(C_{\alpha2} - C_{\alpha1}) = 0.93$. Reading from Fig. 3.21, at the centre of the plate (*i.e.* $y/L = 0$) the corresponding value of Dt/L^2 is equal to about 1.3. The diffusion coefficient for carbon in iron at 450 °C is $3.37 \times 10^{-13}\,\mathrm{m^2/s}$. By substitution the time is equal to 76 s.

(d) How long will it take until the average concentration has fallen to within a factor of two of its final value? This might also be taken to approximate the time required to approach the new equilibrium condition.

We need to determine the time at which the average concentration $\bar{C}(t) = 2C_{\alpha 2}$. In terms of normalized variables this means $[\bar{C}(t) - C_{\alpha 1}]/(C_{\alpha 2} - C_{\alpha 1}) = 0.93$. We can get the result we need from Fig. 3.24. The corresponding value of $(Dt/L^2)^{1/2}$ is given by 0.11, and the time is equal to 64 s. Note the difference between this value and the estimate based on the composition at the centre of the plate. They are quite close.

Earlier we made the assumption that interface motion could be neglected. Is this really valid? The fraction of α-phase in the material is given by the Lever rule, and is equal to $(C_\beta - C_0)/(C_\beta - C_\alpha)$, and so a change of temperature from T_1 to T_2 alters this by the factor $(C_\beta - C_{\alpha 1})/(C_\beta - C_{\alpha 2})$, see Fig. 6.2. This factor will be close to unity under one of two conditions. The first is that the composition of the α-phase does not change very much, *i.e.* $C_{\alpha 1} \approx C_{\alpha 2}$. The second is that the solute content is much larger in the β-phase. This is often the case, and is certainly true for the Fe–Fe$_3$C system. At any rate the width of the diffusion zone, $2L$, is unlikely to change by more than about 10%. As already noted, given the uncertainty in most diffusion data, errors of this order are generally acceptable.

So far we have assumed that the temperature is decreasing; but what is the effect of a temperature increase? Take a new look at the micrographs on the front cover of this book. These show the result of an *in-situ* experiment in which an Al–4 wt% Cu sample is heated within the electron microscope. The CuAl$_2$ particles take the form of plates. Some of these are seen edge-on in the microscope so you can see how thin they are. The two micrographs shown were taken 15 minutes apart and clearly show that the particles have shrunk in size. It should be possible to analyze this experiment based on our previous discussion. If we remain within the two-phase field the effect is similar to that of a temperature decrease, but in the opposite direction. The conditions developed above can be used directly ,only now the solute content in the α-phase increases, as does the relative amount of this phase. I leave it to the reader to see if you can make a quantitative assessment of the process shown on the cover.

But what happens if we increase the temperature above the *solvus*? We consider this next.

6.3   Particle dissolution

6.3.1   Dissolution of plates

We will again consider a periodic array of plates and now choose an alloy of bulk composition C_0, as shown in Fig. 6.5. Since we are only just below the solvus line at the initial temperature T_1, the thickness of the α-plates $2L$, is much larger than that of the β-plates, $2d$. Now suppose that the temperature is increased from T_1 to T_2 in Fig. 6.5, *i.e.* above the solvus line. The final state is clear. The β-plates will dissolve and we will be left with a single α-phase material of uniform composition C_0. But so long as two phases still exist, the composition at the interface is controlled by local equilibrium. Thus the initial and boundary conditions are given by

I.C. $C = C_{\alpha 1}$, $0 < y < L$

B.C. $C = C_{\alpha 2}$, $y = L$

$\partial C / \partial y = 0$, $y = 0$.

Note, however, that the interfacial composition in the α-phase is now greater than that far away from the interface. In other words, at the interface the solute concentration goes from below C_0 to above C_0. There is nothing wrong with this – indeed, thermodynamics dictates that this must be the case. What happens is shown schematically in Fig. 6.6.

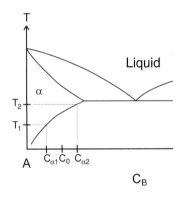

Figure 6.5 An expanded version of the phase shown in Fig. 6.2. This shows the effect on an alloy with bulk composition C_0 of increasing the temperature from T_1 to T_2.

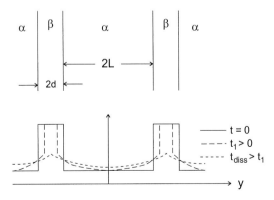

Figure 6.6 A schematic illustration of diffusion profiles during the dissolution of a particle.

Solute diffuses from the β-plates into the α-plates. As it does so, the depletion of solute leads to dissolution of the β-plates, until eventually they disappear. How long will this take? Well, we can again ignore the motion of the interface in answering this. Since the β-plates are thin to begin with and diffusion takes place entirely within the α-phase, the diffusion distance and therefore the kinetics are not greatly affected by interface motion. The time required for dissolution can be obtained by considering the average composition of the α-plates. Initially it is equal to $C_{\alpha 1}$. When dissolution is complete there is no β-phase left and thus the average composition in the α-phase must be C_0. Thus the normalized average composition at this time is given by

$$\frac{\bar{C} - C_{\alpha 1}}{C_{\alpha 2} - C_{\alpha 1}} = \frac{C_0 - C_{\alpha 1}}{C_{\alpha 2} - C_{\alpha 1}}. \tag{6.2}$$

We can therefore determine how much solute is absorbed into the α-plates up to the completion of dissolution and use the solutions for average composition to determine how long this takes. The solution is given in Fig. 3.24 for the planar geometry used here.

Example 6.2: Dissolution of CuAl$_2$ plates in an Al–Cu alloy

Consider the Al-rich end of the Al–Cu phase diagram. Suppose that an alloy containing 3 wt% Cu has been aged at 150 °C so as to develop semi-coherent plates of the CuAl$_2$ phase. These plates are 10 nm thick, with an average separation of 330 nm. The temperature is now increased to 520 °C.

(a) What are the initial and boundary conditions for this problem?

These can be read directly from the phase diagram (see Appendix C), assuming local equilibrium at the interface between the Al-rich α-phase and the CuAl$_2$ phase. They are:

I.C.	$X^* = X^*_{\alpha 1} = 0.5\,\text{wt}\%,$	$0 < y < L$
B.C.	$X^* = X^*_{\alpha 2} = 5.0\,\text{wt}\%,$	$y = y_1$
	$\partial X^*/\partial y = 0,$	$y = 0,$

where y_1 denotes the current position of the α/β interface. From the statement of the problem the width of the α-phase region $2L$, is 320 nm.

(b) How long will it take for the CuAl$_2$ plates to completely dissolve?

The plates will have dissolved once the average composition in the α-phase is equal to the bulk composition of the alloy. Thus the normalized average composition is

$$\frac{\bar{C} - C_{\alpha 1}}{C_{\alpha 2} - C_{\alpha 1}} = \frac{\bar{X}^* - X^*_{\alpha 1}}{X^*_{\alpha 2} - X^*_{\alpha 1}} = 0.56.$$

Reading from Fig. 3.24 for the normalized time, $(Dt/L^2)^{1/2}$ is equal to 0.48. Because the α-phase is a dilute solution of Cu in Al (maximum solubility is 5 wt%), the appropriate diffusion coefficient is that for Cu in Al at 520 °C. This can be determined from the tabulation in Appendix B. It is

$7.32 \times 10^{-14}\,\mathrm{m^2/s}$. Thus the time to dissolve the plates is only $0.08\,\mathrm{s}$. In other words, because of the short diffusion distance involved, the $CuAl_2$ will dissolve almost instantaneously upon heating.

What happens after the plates are completely dissolved? As shown in Fig. 6.6, once this occurs there is still a substantial concentration gradient in the α-phase. But we no longer have an interface to establish a condition of local equilibrium. The problem now becomes one of homogenization of the single-phase material. The new boundary conditions become

$$\text{B.C.} \quad \partial C/\partial y = 0, \qquad y = L + d$$
$$\partial C/\partial y = 0, \qquad y = 0.$$

The initial conditions for this problem are given by the composition profile in the α-phase at the completion of dissolution. We can determine this from solutions to the problem we just solved but the exact solution is mathematically complex. It is much simpler, however, to approximate this as a sinusoidal profile

$$C(y, t_{\mathrm{diss}}) = C_0 + (C_0 - C_{\alpha 2}) \cos\left(\frac{\pi y}{L + d}\right). \tag{6.3}$$

Here t_{diss} represents the time at which particle dissolution occurs. We developed the solution for homogenization with this kind of initial condition in an earlier chapter (see eq. (3.52)).

This is another example of a diffusion problem in which the phases present can change, and thus the boundary conditions are altered. We need to be aware of this, and divide the problem into segments, each of which can be modelled separately.

6.3.2 Dissolution of spherical particles

In many materials second-phase particles are more or less spherical. How does the dissolution of these particles differ from that of plates? At first sight this looks like a straight-forward substitution of spherical for planar symmetry in the previous problem. However, it is not that simple. Consider a random array of β-particles in an α-matrix as shown in Fig. 6.7. The particles have radius R and are separated by an average distance of $2L$. In order to model such a complex geometry we use a technique common in many fields of physical modelling – we develop a 'unit cell' which represents the key features of the structure. In this case, a unit cell would consist of a single β-particle, surrounded by a spherical region of radius L. The outer boundary of this region is therefore about half-way from the particle of interest to all of its neighbours. We can assume that at this boundary the flux is small, i.e. that this is a symmetry boundary. We will again assume that the temperature is raised from T_1 to T_2 as illustrated in Fig. 6.5. We can now write down a set of initial and boundary conditions that describe this problem:

I.C. $C = C_{\alpha 1}$, $R < r < L$
B.C. $C = C_{\alpha 2}$, $r = R$
          $\partial C / \partial r = 0$, $r = L$.

This is just a revised version of the conditions obtained for plates, but with spherical symmetry. So we can use the solutions developed in Chapter 3 for this symmetry, right? Wrong!!! Think carefully about those solutions for spherical symmetry. We mostly developed solutions for diffusion into or out of a **solid** sphere. In that case, one of the boundary conditions is always $\partial C / \partial r = 0$ at $r = 0$. This is clearly different from the problem to be solved here in which the sphere is hollow, *i.e.* the spherical α-phase region contains a β-phase particle. This problem is close to that of a hollow sphere with constant boundary conditions on the inner and outer surface (Section 3.6.4). For the case where the boundary concentrations were equal we saw that the solution approximated that of a plate with the thickness equal to the wall thickness of the sphere. Because we have a symmetry condition at the outer boundary of the sphere ($r = L$) in this case, we can get an approximate solution from that of a plate of thickness $2(L - d)$ with $C = C_{\alpha 2}$ at each surface.

One of the keys to this problem is estimating a value for L. The most logical way to do this is to let the volume fraction of particles within the sphere of radius L equal the volume fraction of the material as a whole, φ, which we can determine from the phase diagram. From geometry you can show that

$$\varphi = \left(\frac{R}{L}\right)^3 \quad \text{or} \quad L = \frac{R}{\varphi^{1/3}}. \tag{6.4}$$

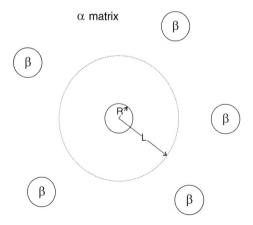

Figure 6.7 A series of β-particles in equilibrium with the α-matrix. The average separation between the particles is $2L$.

Example 6.3: Dissolution of spherical CuAl$_2$ particles in an Al–Cu alloy

This is a modified version of the previous example which concerned an Al alloy containing 3 wt% Cu. We now consider spherical particles, with the same separation, 330 nm, equilibrated at 520 °C.

(a) What is the average diameter of the particles?

From eq. (6.4) we know that $R = L \cdot \varphi^{1/3}$. From the phase diagram (see Appendix C) we find that $X^*_{\alpha 1} = 0.5$ wt%, while $X^*_{CuAl_2} = 53$ wt%. Using the Lever Rule, the weight fraction of CuAl$_2$ particles is

$$f^*_{CuAl_2} = \frac{\bar{X}^* - X^*_{\alpha 1}}{X^*_{CuAl_2} - X^*_{\alpha 1}} = \frac{2.5}{52.5} = 4.7 \, \text{wt\%}.$$

This is converted to a volume fraction through the densities of the individual phases, according to the relationship given in Table 2.1, p. 30. Thus

$$\varphi = f^* \frac{\rho_{alloy}}{\rho_{CuAl_2}} = \frac{f^*}{\rho_{CuAl_2}} \left(\frac{f^*}{\rho_{CuAl_2}} + \frac{1 - f^*}{\rho_{Al}} \right)^{-1}.$$

The density of Al is 2.7 g/cm^3, while that of CuAl$_2$ is 4.34 g/cm^3. Therefore,

$$\rho = \frac{0.047}{4.34} \left(\frac{0.047}{4.34} + \frac{0.953}{2.7} \right)^{-1} = 2.98 \, \text{wt\%}.$$

We can now determine the particle diameter to be

$$2R = 2L \cdot \varphi^{1/3} = 330 \times (0.0298)^{1/3} = 103 \, \text{nm}.$$

(b) How long will it take these particles to dissolve at 520 °C?

This part of the problem is virtually identical to the previous example. In Section 3.6.4 we learned that the average concentration in a hollow sphere was similar to that in a plate after an appropriate conversion of the normalized time (see Fig. 3.28). So we will treat this case as a plate, but with a thickness, $2(L - R) = 227$ nm. The normalized time for dissolution, $Dt/(L - R)^2$, is 0.48, as in the previous example. The diffusion coefficient is 7.32×10^{-14} m^2/s. Therefore the time to dissolve the particles is reduced by $(L - R)^2/L^2 = 0.47$ to 0.04 s.

6.4   Growth of second-phase particles by solute rejection

6.4.1   Planar growth

We now wish to consider a problem in which motion of the interface *is* important – the growth of precipitates. In fact, we will attempt to determine how fast the interface moves as the precipitate grows. Consider a phase diagram such as that shown in Fig. 6.8. There are three solid-phase regions, labelled α, β and γ. A second phase will be precipitated from a single-phase matrix if we lower the temperature either from the γ-regime into the $(\alpha + \gamma)$-regime, or from the α-regime into the $(\alpha + \beta)$-regime. We will consider the first case but the development is identical for each. A practical example of this type is the growth of ferrite from austenite in plain carbon steels. The precipitation processes which occur during age hardening of alloys are mostly examples of the second type. Figure 6.8 is representative of the relevant portion of the Fe–C phase diagram (although the composition of the β-phase, *i.e.* Fe$_3$C) does not change with temperature in that case. Ferrite grows primarily in two different morphologies. At higher temperatures it grows with a planar growth front from the austenite grain boundaries (so-called 'chunky' ferrite). At lower temperatures a spikey growth morphology called Widmänstatten ferrite is observed.[¶]

The growth of chunky ferrite is planar and can therefore be readily modelled. So, let us consider the growth of a ferrite particle with a planar front, starting from an austenite grain boundary, as shown schematically in Fig. 6.9. Suppose that the bulk composition of the alloy is C_0 (see Fig. 6.8), and that we quickly lower the temperature from within the γ-regime at temperature T_1 to a temperature T_2 within the two-phase field. We wish to determine the rate of movement of the interface. This will be limited either by the rate of reaction at the interface or by the rate of diffusion of solute into the parent phase (in this case the γ-phase). If the interface is able to maintain a local equilibrium condition, as we have generally assumed, then the rate of diffusion controls the kinetics of this process. As the α-phase grows it maintains a concentration C_α, as noted on the phase diagram in Fig. 6.8.[†] Since the solute content C_α of the precipitating phase is less than that of the original phase, we say that the

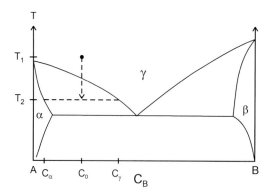

Figure 6.8 A schematic of part of a phase diagram exhibiting a eutectoid reaction.

[¶] For a more complete description of these morphologies and the conditions under which they grow see P. G. Shewmon, *Transformation in Metals* (McGraw-Hill, New York; 1969), pp. 216–19.

[†] This is different from the case of solidification, in which the evolution of heat (the latent heat of solidification) slows the cooling process. There, we precipitate the α-phase with a range of composition. The heat evolved during solid phase transitions is much smaller, enabling the process to occur isothermally.

precipitate grows by *solute rejection*, *i.e.* it can only grow by moving the solute away from the interface into the original phase. The composition of the γ-phase in equilibrium with the precipitate at T_2 is C_γ. After some growth has occurred a concentration profile such as that shown in Fig. 6.9 must develop. Since the bulk composition is constant and equal to C_0, the area below C_0 on this profile (which represents the amount of solute rejected by the growing α-phase), must equal the area above C_0 (which represents the amount of solute redistributed into the γ-phase). It is easy to see that as the precipitate grows, rejecting ever more solute, the distance that the newly rejected solute will have to travel away from the interface continues to increase. Thus we might expect that the rate at which a precipitate grows will decrease with time. We will now determine how fast it decreases.

Let us begin by writing down the boundary conditions for this problem. Given the preamble above they can be simply stated as:

I.C. $C = C_0$, $y \geq 0$
B.C. $C = C_\gamma$, $y = y_{\mathrm{I}}$
        $C = C_0$, $y \to \infty$.

These conditions are applicable at short times. At long times we can replace the condition at infinity by a symmetry condition ($\partial C / \partial y = 0$) at, say, half the grain size of the γ-phase. In either case, this set of boundary conditions does not completely define the problem since the position of the interface $y = y_{\mathrm{I}}$ is unknown and is not constant. We therefore need another condition which tells us something about how the interface moves.

This comes from the condition of conservation of mass. As the interface moves by an infinitesimal amount $\mathrm{d}y_{\mathrm{I}}$, a fixed amount of solute must be removed from the interface. As you can see in Fig. 6.9, the concentration

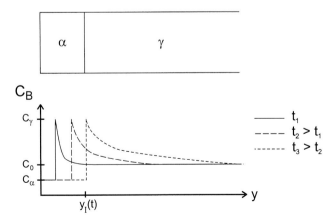

Figure 6.9 The diffusion profiles ahead of a precipitate growing by solute rejection. As the precipitate grows the solute must diffuse ever further from the interface.

drops from C_γ to C_α as the interface passes. Therefore, the amount of solute removed is equal to the amount of solute that was in this region when it was γ-phase, i.e. $C_\gamma \cdot A dy_I$, less the amount left over after it has been converted to α-phase, i.e. $C_\alpha \cdot A \, dy_I$. This is given by $(C_\gamma - C_\alpha)A \cdot dy_I$. Here A is the cross-sectional area of the moving interface.

All of this solute must be removed from the interface by diffusion in the γ-phase. Let us suppose that the interface moves by dy_I in a time dt. Then the amount of solute diffusing away from the interface is equal to $J \cdot A \cdot dt$. We can equate these two terms to determine the relationship between the diffusive flux at the interface and the rate of interface motion. The result is

$$\frac{dy_I}{dt} = \frac{J}{C_\gamma - C_\alpha}. \tag{6.5}$$

Well this is all very fine. We now have an equation for interface motion. It tells us that this depends on the rate of diffusion. But the boundary conditions we developed earlier tell us that the rate of diffusion depends on the interface position. This means that the two problems of diffusion and interface motion are coupled, i.e. they must be solved simultaneously. This can be done but it makes the mathematics considerably more complex. Is there any way we can simplify the problem and still arrive at a useful solution? The answer is yes. Let us suppose that the rate of interface motion is relatively slow, such that the distance travelled by the interface is small compared to the distance that atoms have to diffuse away from the interface. If this happens then the diffusion profile is not strongly affected by the movement of the interface. Let us examine the equation just developed for interface motion more carefully. It contains the flux J at the interface. If we substitute Fick's First Law here we get

$$\frac{dy_I}{dt} = \frac{D}{C_\gamma - C_\alpha} \left(\frac{\partial C}{\partial y} \right)_{y=y_I}. \tag{6.6}$$

If interface motion is sufficiently slow then the concentration gradient can be estimated from the solutions we have already developed for diffusion from a surface (or interface) of fixed concentration. This is the familiar error function solution

$$\frac{C(y,t) - C_0}{C_\gamma - C_0} = \mathrm{erfc}\left(\frac{y - y_I}{2\sqrt{Dt}} \right). \tag{6.7}$$

Note that we have shifted the origin of this solution to the interface at $y = y_I$. We can get an estimate of the slope of this function at the interface very easily. Remember that we determined the meaning of $\sqrt{Dt}$ by noting that the error function drops to half its surface value over a distance of $\sqrt{Dt}$. Therefore the slope must be about $(C_0 - C_\gamma)/(2\sqrt{Dt})$. A more exact estimate can be made by differentiation of the error function equation. We find (see box below) a very similar result, $(C_0 - C_\gamma)/\sqrt{\pi D t}$. We can now substitute this back into the equation for interface motion. The result is

$$\frac{\mathrm{d}y_\mathrm{I}}{\mathrm{d}t} = +\frac{C_\gamma - C_0}{C_\gamma - C_\alpha} \sqrt{\frac{D}{\pi t}}. \tag{6.8}$$

This solution is reminiscent of those we developed for the pseudo-steady-state growth of films (such as by oxidation). The interface motion is governed in both cases by parabolic kinetics, *i.e.* the rate decreases as $t^{-1/2}$. The reason is simply that the distance over which diffusion must occur increases with time.

Solution based on differentiating the error function

The exact derivation of the concentration gradient at the surface requires that we differentiate the error function. Recall that the error function complement is defined as

$$\mathrm{erfc}(z) = \frac{2}{\sqrt{\pi}} \int_z^\infty \mathrm{e}^{\eta^2} \, \mathrm{d}\eta.$$

The derivative of this at any value of z is therefore

$$\frac{\partial \mathrm{erfc}(z)}{\partial z} = -\frac{2}{\sqrt{\pi}} \mathrm{e}^{z^2}.$$

In order to relate this to the diffusion profile ahead of an interface we have to recognize that z is the normalized distance ahead of the interface given by $(y - y_\mathrm{I})/(2\sqrt{D t})$. The concentration gradient is related to the normalized variable solution such that

$$\frac{\partial\left(\dfrac{C(y, t) - C_0}{C_\gamma - C_0}\right)}{\partial y} = \frac{1}{C_\gamma - C_0} \frac{\partial C(y, t)}{\partial y},$$

so that (see eq. (6.7))

$$\frac{\partial C(y, t)}{\partial y} = (C_\gamma - C_0) \frac{\partial \mathrm{erfc}(z)}{\partial z} \frac{\partial z}{\partial y},$$

and thus

$$\frac{\partial C(y_\mathrm{I}, t)}{\partial y} = -\frac{2}{\sqrt{\pi}} \frac{1}{2\sqrt{D t}} (C_\gamma - C_0)$$

$$= \frac{C_0 - C_\gamma}{\sqrt{\pi D t}}.$$

Before proceeding, we should check our assumption regarding the slow rate of interface motion with respect to diffusion distances. This is equivalent to requiring that y_I is small with respect to $\sqrt{Dt}$, or that $\partial y_I / \partial t$ is small with respect to $\sqrt{D/t}$. From the solution we have just developed (eq. (6.8)), it is clear that this condition is met provided that $(C_\gamma - C_0)/(C_\gamma - C_\alpha)$ is much less than $\sqrt{\pi}$. We note that $(C_\gamma - C_0)/(C_\gamma - C_\alpha)$ is always less than unity, so this assumption will generally be valid. However, it is most closely obeyed for small undercoolings, when C_γ is close to C_0.

6.4.2   Precipitation of spherical particles

In many materials precipitates do not grow on planar fronts but rather as spheres. If the diffusion distance is significant with respect to the particle diameter we might expect this will affect the kinetics of precipitation. So what must we do to modify the previous solution for spherical particles?

We will maintain the same assumptions, in particular that the particle grows slowly with respect to diffusion into the matrix. The initial and boundary conditions are unchanged except for the change of variables to spherical coordinates. Thus

$$
\begin{array}{lll}
\text{I.C.} & C = C_0, & r \geq R \\
\text{B.C.} & C = C_\gamma, & r = R \\
& C = C_0, & r \to \infty.
\end{array}
$$

where R is the particle radius. The change of geometry does not affect the mass conservation relationship (eq. (6.5)) except that y_I is replaced by R. But the diffusion profile is different. Fick's Second Law for spherical symmetry is given by eq. (3.11):

$$
\frac{\partial C}{\partial t} = D\left(\frac{\partial^2 C}{\partial r^2} + \frac{2}{r}\frac{\partial C}{\partial r}\right).
\tag{6.9}
$$

Now to simplify this problem we can use an approximation that works best for spherical coordinates. Because the cross-sectional area through which solute diffuses expands rapidly as r increases, spherical problems closely approximate pseudo-steady-state conditions, *i.e.* the concentration profile away from the surface of the sphere is more or less constant. Thus, $\partial C/\partial t \approx 0$ and the above equation has the solution

$$
C = C_0 - \frac{R}{r}(C_0 - C_\gamma).
\tag{6.10}
$$

From this we can determine the flux at the particle surface. It is

$$J = -D \left.\frac{\partial C}{\partial r}\right|_{r=R} = D\,\frac{C_\gamma - C_0}{R}. \tag{6.11}$$

We can now substitute this back into eq. (6.5) to get the particle growth rate

$$\frac{\mathrm{d}R}{\mathrm{d}t} = \frac{C_\gamma - C_0}{C_\gamma - C_\alpha}\,\frac{D}{R}. \tag{6.12}$$

This result looks rather different from that which we derived for a planar growth morphology (eq. (6.8)). However, if we integrate each result to determine the particle size as a function of time the results are actually quite similar (as shown in the following example).

Example 6.4: Comparison of growth kinetics for planar and spherical precipitates

Compare the time to grow a plate-like particle with that of a spherical particle. Assume that the initial size of each particle (width or diameter) is 2 nm and you wish to know the time to grow each to 10 nm in size. Suppose that the values of C_α, C_0 and C_γ are 1, 5 and 10% respectively.

To solve this problem we need to integrate the growth rates. For the planar case, the integration of eq. (6.8) follows directly from other problems involving parabolic kinetics. Thus

$$L - L_0 = 2\,\frac{C_\gamma - C_0}{C_\gamma - C_\alpha}\,\sqrt{\frac{Dt}{\pi}},$$

where L_0 and L are the initial and final half-widths of the precipitate plates. In terms of time, therefore,

$$t_\mathrm{p} = \frac{\pi}{4D}\left[\frac{C_\gamma - C_\alpha}{C_\gamma - C_0}\,(L - L_0)\right]^2.$$

For spherical particles, integration of eq. (6.12) gives

$$t_\mathrm{s} = \frac{1}{2D}\,\frac{C_\gamma - C_\alpha}{C_\gamma - C_0}\,(R^2 - R_0^2).$$

The relative time, as given by the ratio of t_p to t_s, is

$$\frac{t_\mathrm{p}}{t_\mathrm{s}} = \frac{\pi}{2}\,\frac{C_\gamma - C_\alpha}{C_\gamma - C_0}\,\frac{(L - L_0)^2}{R^2 - R_0^2}.$$

For the parameters given above, this ratio is 2.35, *i.e.* it takes a bit more than twice as long to grow the planar particle. As the size of the precipitates increases such that L and R are large with respect to L_0 and R_0 respectively then this ratio increases to 3.53.

Of course one could phrase the question in a different (and perhaps fairer) way, by asking how long it takes precipitates to grow to an equivalent volume. This is left to the reader.

6.5   Cooperative growth processes

Consider once again the schematic phase diagram shown in Fig. 6.8. This diagram includes a eutectoid reaction in which a single high-temperature solid phase (γ) decomposes into two different solid phases (α and β). Similar reactions are often found during solidification (the eutectic reaction), except that the starting phase is a liquid. The growth of two phases from a high-temperature phase (whether solid or liquid) involves 'cooperative' growth. The most common example is the pearlite reaction in plain carbon steel, as shown in Fig. 6.4(a). This reaction is shown schematically in Fig. 6.10. The two phases which grow develop side-by-side. Thus the solute rejected from one phase is absorbed by the other, a process called solute 'partitioning'. A variety of growth morphologies are possible such as plates, rods or discrete particles. The last of these is not very common since it requires that one of the phases is repeatedly re-nucleated. However, both plates and rods are often found. We will consider plate-like growth because the two-dimensional geometry is easier to treat. But the solution for rod-like growth will be very similar.

In order to proceed we need to think about the process of cooperative growth in more detail. If we refer to the phase diagram (Fig. 6.8) then we see that the solute, call it B, is partitioned from the α-phase to the β-phase. What this means is that far away from the interface the solute concentration in the γ-phase is uniform. But near the interface the solute diffuses towards the region just ahead of the solute-rich (β) phase. Conversely, the solute diffuses away from the interface ahead of the solute-poor (α) phase. This is illustrated in Fig. 6.11. Since we are interested in how fast the solute gets from the α- to the β-phase the important direction for diffusion is actually parallel to the interface, in the z-direction. This diffusion is driven by the fact that the local equilibrium concentration in the γ-phase is different just ahead of a growing α-plate than it is just ahead of a β-plate. Now why is this so and how can we determine the exact compositions?

In order to answer this we need to learn something about phase diagrams which you may or may not have learned when you were taught thermodynamics. We are interested in the cooperative growth of a eutectoid phase

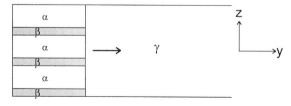

Figure 6.10 A schematic illustration of precipitation due to an eutectoid reaction. This form of precipitation is called 'cooperative growth'.

at some temperature below the eutectoid temperature. Why is this? Well, even during solidification things happen very slowly at the melting point. With solid–solid phase transformations this is especially so. The driving force for phase transformations goes to zero at the equilibrium condition. We therefore need a certain amount of undercooling to drive the transformation at a reasonable rate. Let us suppose then that a eutectoid transformation is proceeding at a temperature T^* on the phase diagram (Fig. 6.12). An ordinary phase diagram does not show the equilibrium composition of the γ-phase below the eutectoid temperature since the phase is not stable there. But we can estimate this by extrapolating the γ solubility limits from above the eutectoid. Why is this possible?

To understand this we need to refer to free-energy curves of the type used to explain and calculate phase diagrams. Three sets of such curves are shown in Fig. 6.13. The top set refers to a temperature above the eutectoid. All three phases are stable and the common tangent construction shows the solubility limits of each phase. The range of composition exhibited by the γ-phase is between $C_{\gamma\alpha}$ and $C_{\gamma\beta}$. The middle set of curves refers to the eutectoid

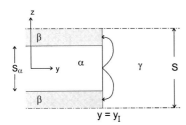

Figure 6.11 When a two-phase composite such as pearlite is precipitated by cooperative growth, the diffusion is primarily parallel to the interface. The diffusion distance is a constant equal to the periodicity of the plates.

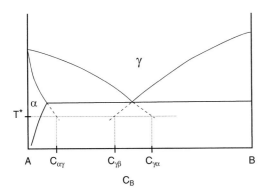

Figure 6.12 The condition of local equilibrium at internal interfaces at temperature T^* is obtained by extrapolating the lines on the phase diagram beyond their nominal phase-field boundaries.

temperature. All three phases are still stable, but a single line is tangent to all three curves. The γ-phase is now stable only at a single composition. The bottom set of curves refers to a temperature below the eutectoid. The common tangent construction now shows that only the α- and β-phases are actually stable. However, if *locally* we have α- and γ-phases present then the common tangent between these two phases still gives the *local* equilibrium concentrations. You can see that at this temperature γ-phase in equilibrium with α has a

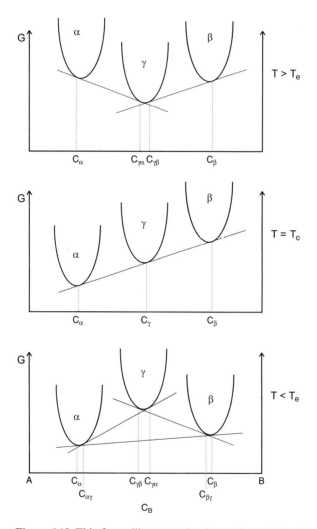

Figure 6.13 This figure illustrates the thermodynamic justification for extrapolating lines beyond their normal phase-field boundaries, as shown in the previous figure. The phase boundaries are constructed from the common tangent between fields. When two phases are in contact, they will form a local equilibrium. Such is the case for the γ-phase in contact with both the α- and β-phases at temperature below the eutectoid temperature T_e.

higher solute content than γ-phase in equilibrium with β. (Contrast the positions of $C_{\gamma\alpha}$ and $C_{\gamma\beta}$ on the top and bottom diagrams of Fig. 6.13. They are reversed.) If we consider the equilibrium between the γ- and α-phases alone we can see that nothing magical happens at the eutectoid temperature. The solubility of the two phases continues to change as the relative depth of the free-energy wells change with temperature. Thus we can reasonably estimate these solubilities by extrapolation of lines on the phase diagram. These extrapolation are shown as dashed lines on Fig. 6.12.

At the temperature of interest, T^*, the local equilibrium concentrations can be read from the phase diagram (Fig. 6.12). Since we have determined that diffusion in the γ-phase (between the γ/α and γ/β interfaces) is what controls the process, the boundary conditions are given by

$$\text{B.C.} \qquad C = C_{\gamma\alpha}, \qquad z = 0, \qquad y = y_{\mathrm{I}}$$
$$C = C_{\gamma\beta}, \qquad z = \tfrac{1}{2}S, \quad y = y_{\mathrm{I}}.$$

Here, S is the periodicity of the plates (see Fig. 6.11). This is determined by a range of factors including the temperature and the free energy of the α/β interface. The diffusion distance is rather short, about $\frac{1}{2}S$. Therefore we can expect that a steady state will be established rather quickly. The flux between the α- and β-plates is therefore given by Fick's First Law with the concentration gradient equal to $\partial C/\partial z = 2(C_{\gamma\beta} - C_{\gamma\alpha})/S$.

To determine the rate of interface motion we again need to perform a mass balance. As the interface moves by an amount dy_{I} the amount of solute which is partitioned is equal to $(C_{\gamma\alpha} - C_\alpha) \cdot L \cdot S_\alpha \cdot dy_{\mathrm{I}}$, where S_α is the thickness of the α-plates only and L is the length of the diffusing interface (see Fig. 6.11). A similar result is obtained if we do the calculation for the β-plates. As before, we equate this with the amount of solute that can diffuse in time dt, which is $J \cdot A \cdot dt$. Remember that A is the cross-sectional area of the diffusive path and here we have to make an approximation. We do not exactly know how wide the path is. Solute diffuses away from one interface and loops around toward the other. The width of the loop is uncertain but it is reasonable to assume it will be about as wide as it is long, *i.e.* A is approximately equal to $L \cdot S_\alpha$. When we equate these two terms we get as the rate of interface motion:

$$\frac{dy_{\mathrm{I}}}{dt} = \frac{C_{\gamma\beta} - C_{\gamma\alpha}}{C_{\gamma\alpha} - C_\alpha} \frac{2D}{S}. \qquad (6.13)$$

You should note that this is rather similar in form to the solution for precipitate growth by solute rejection except for one important factor. Whereas before the diffusion distance was $\sqrt{Dt}$ and thus precipitate growth slowed down over time, here the diffusion distance is a constant, equal to the plate spacing S. This is because in cooperative growth there is no requirement for long-range diffusion. This makes it a very efficient growth process.

6.6   **Further reading**

P. G. Shewmon, *Transformations in Metals* (McGraw-Hill, New York, 1969)

D. A. Porter and K. E. Easterling, *Phase Transformations in Metals and* *Alloys* (Van Nostrand Reinhold, Wokingham, UK 1981)

J. S. Kirkaldy and D. J. Young, *Diffusion in the Condensed State* (The Institute of Metals, London, 1987)

6.7   **Problems to chapter 6**

Note: Relevant phase diagrams are to be found in Appendix C.

6.1   In each of the situations (a)–(g) below assume that the system is originally in equilibrium. The indicated change is made to the system. In each case:

(i)   Plot a curve (which may be a straight line) for the initial conditions. Indicate how you arrive at your numerical value.

(ii) Draw a curve for infinite time (*i.e.* the final state following the change of conditions), and indicate whether it is a steady-state or an equilibrium condition.

(iii) Give the boundary conditions which apply to each situation.

(iv) Draw a set of schematic curves for increasing time (three or four curves will probably suffice), which show how the composition varies with position and time.

Clearly state any assumptions you make!

(a)   A block of aluminum is attached to a block of silicon at 575 °C. The temperature is then reduced to 500 °C.

(b)   A periodic structure of alternating plates of aluminum and silicon (pearlitic structure) is at 575 °C. The temperature is decreased to 500 °C.

(c)   A block of MgO is attached to a thin slice of $MgCr_2O_4$ at 1000 °C. The temperature is increased to 1600 °C.

(d)   A slab of CoO (a p-type oxide with excess oxygen) sits in an atmosphere containing 0.2 atm. of oxygen. The oxygen pressure is increased to 1 atm.

(e)   A piece of clay is saturated with water. It is then placed in a sealed jar of dry air.

(f)   The same piece of clay, having been left in the sealed jar for a long time, is then placed in a large oven at 50 °C with 0.01 atm. water vapour.

(g) A block of ZrO_2–30 wt% CaO is held at 2250 °C. The temperature is decreased to 1500 °C.

6.2   Suppose that a Cu–37.5 wt% Zn alloy is equilibrated at 800 °C. Suppose the microstructure consists of an α-matrix containing (nominally) spherical β-particles spaced 100 nm apart. The temperature is now lowered to 700 °C.

(a) Describe (as precisely as possible) the boundary-value problem you would need to solve to determine how long the system will take to re-equilibrate.

(b) Draw a set of schematic curves indicating what will occur. Do not assume that the interfaces are stationary. Do assume that diffusion is more rapid in the β-phase.

(c) If the process is controlled by diffusion in the α-phase, estimate how long it will take to reach a new equilibrium.

6.3   An Al–3 wt% Cu alloy has been heat treated at 300 °C to precipitate coarse particles of θ-phase, with an average particle diameter of 0.25 µm.

(a) If the alloy is at equilibrium, estimate the weight fraction of particles in the alloy, and their average separation.

(b) Suppose you now anneal that alloy at 550 °C. Assuming a planar geometry, estimate how long it will take until the particles have dissolved.

6.4   Copper is added to aluminum in order to improve strength through an age-hardening process. Consider an alloy containing 4 wt% Cu. This material is first annealed at 550 °C in order to develop a solid solution. It is then quenched to room temperature, at which point small particles are nucleated with an average separation of 100 nm. It is finally aged at 175 °C to allow the precipitates to grow.

(a) What is the initial condition (in terms of copper concentration) at the start of the aging process?

(b) What boundary conditions apply during aging?

(c) Draw a set of schematic diagrams which illustrate how the copper concentration changes with time during aging. Be as precise as possible, label concentrations properly and draw the curves with care.

(d) How long will it take until the precipitation process is complete, as defined by when the average concentration in the matrix is within 10% of (*i.e.* 1.1 times) its final value?

Note: Since the phase diagram only goes down to 300 °C, assume that the concentrations at 175 °C are the same as those as 300 °C.

6.5   An alloy of Cu plus 70 wt% Zn is annealed at 550 °C until equilibrium is achieved. It is then cooled to 400 °C and held for some time. Assume that the new phase forms as plates 100 nm apart.

(a) Draw a set of properly labelled schematic curves to show what will happen. State any assumptions you make.

(b) Calculate the rate of motion of the interface after 1 hour. Estimate the average composition of the γ-phase at this time.

(c) What will be the final thickness of the plates?

(d) In estimating how the plates grow with time, you need to make some assumptions about the nature of the diffusion profile ahead of the growing plates. What is this assumption? Use the assumption to estimate the thickness of the plates when the diffusion profiles from neighbouring plates begin to overlap.

Use $D = 10^{-19}\,\mathrm{m^2/s}$ in the γ-phase.

6.6   Consider the Al–Zn phase diagram and alloys of the following compositions: (i) Al–25 wt% Zn; (ii) Al–78 wt% Zn.

Suppose that each alloy is annealed at 350 °C and then cooled to 200 °C.

(a) Develop a model for the process which will take place in the Al–78 wt% Zn alloy upon cooling. You may assume that nucleation is complete, and you may develop a one-dimensional model.

(b) Briefly indicate how the reaction differs in the other alloy.

(c) How does the rate of precipitate growth depend on time for the two cases?

6.7   Suppose that a piece of antimonial copper consisting of Cu plus 5 wt% Sb has been heat treated at 250 °C so as to give large plate-like particles 10 nm thick, aligned in a single direction. This material is now annealed at 390 °C.

(a) Write down the initial and boundary conditions, being careful to state your assumptions. Reproduce the appropriate part of the phase diagram and label it with the important compositions.

(b) Draw a set of schematic curves for the concentration of Sb at different times from $t = 0$ to $t = \infty$.

(c) Determine the length of time required at 390 °C before the alloy becomes single phase.

6.8   Suppose that a cast Al–1 wt% Si alloy, contains equilibrium Si particles 1 μm in diameter. This alloy is to be heat treated by first putting all of the Si particles back into solution.

(a) Draw a set of schematic curves indicating what will happen during this process.

(b) Indicate the initial conditions, the final conditions (once a new equilibrium is reached) and the boundary conditions during dissolution.

(c) Estimate how long the alloy should be held at 550 °C so as to ensure that all of the Si is dissolved.

6.9   Suppose that a Cu–2.5 wt% Mg alloy is annealed for a long period of
        time at 700 °C. The temperature is then lowered to 300 °C. The pre-
        cipitates which form have an average centre-to-centre spacing of
        200 nm.
        (a) Describe as precisely as possible, the initial and boundary condi-
            tions you would specify, in order to determine how long the sys-
            tem will take to come to equilibrium.
        (b) Draw a set of schematic curves showing what will occur. Label
            these curves are precisely as possible.
        (c) If the diffusion coefficient for Mg in Cu at 300 °C is 4×10^{-18} m^2/s,
            determine how long it will take until the average Mg concentration
            in the matrix drops to 2 wt%. (*Note:* You may assume that the
            diffusion of Mg in the matrix controls the process.)
6.10   Suppose that the surface of a pure iron plate is treated with chromium
        at 1000 °C, so as to maintain a concentration at the surface equal to
        50 wt% Cr. The diffusivity of Cr is much higher in α-Fe than in γ-Fe.
        (a) Draw a set of precisely labelled schematic curves for the Cr con-
            centration with depth at different times. Pay attention to the
            slopes and curvatures in your drawings. You will need to think
            carefully about what the phase diagram tells you.
        (b) Derive an expression for the depth of the surface phase as a function
            of time. Be sure to state and justify any assumptions you make.
6.11   Suppose that a Cu–Co alloy containing 6 wt% Co is annealed for a
        long time at 1050 °C, and then cooled to 700 °C. This nucleates Co-
        rich particles with an average spacing of 500 nm. You wish to model
        the process by which these particles grow.
        (a) Draw a picture of the process indicating how you will model it,
            e.g. what geometry will you use?
        (b) Write down the initial and boundary conditions.
        (c) Draw a set of schematic curves, labelled as precisely as possible,
            for the concentration profile at different times. Do not assume
            that the interface is stationary.
        (d) Estimate how long it will take to reach equilibrium at this tem-
            perature.
6.12   An industrial problem of considerable interest in the steel industry
        involves the austenitization of steel, *i.e.* the process whereby a steel
        consisting of ferrite and cementite is converted into austenite.
        Consider a steel of eutectoid composition (0.8 wt% C) which consists
        entirely of pearlite. The steel has been cold-rolled so that the grains are
        pancake-shaped, approximately 25 μm thick by 100 μm wide and
        250 μm long. The steel is annealed at 800 °C. Austenite is assumed
        to nucleate uniformly along the grain boundaries.

(a) Develop a model for the rate of growth of austenite into the pearlite at this temperature. Be sure to show what you are doing and to illustrate your model with schematic diagrams. Derive the model from first principles.

(b) Use your model to estimate the time required for complete conversion of the original pearlite structure to austenite.

Assume that the pearlite structure is randomly oriented with a spacing between the ferrite plates of 5 μm.

6.13   Many Ni-base superalloys are based on Ni–Al alloys. If an alloy Ni + 7 wt% Al is produced it can be heat treated to produce a fine dispersion of γ'-phase (Ni_3Al) particles in the Ni-rich phase (called the γ-phase). During service the γ'-phase particles coarsen (*i.e.* the average particle size increases) which degrades the material's properties. The material can be 'rejuvenated' by annealing at 1150 °C to dissolve the particles, then repeating the original heat treatment procedure. Suppose that at the start of this procedure the particles (which are cubic in shape) are 90 nm in diameter. Assume that Ni diffusion in the Ni-rich phase is rate controlling.

(a) Write down the initial and boundary conditions.

(b) Draw (carefully, and properly labelled) schematic curves, indicating how the concentration will change with time.

(c) Make a rough estimate of the time required to dissolve the particles, assuming you can raise the temperature quickly enough to 1150 °C.

(d) What would be the interface velocity at 1% of the time you calculated in part (c).

6.14   A plain carbon steel with 0.8 wt% (4 at.%) C consists of a pure pearlite structure. If this steel is now annealed at 730 °C in a C-free atmosphere, a layer of pure ferrite develops on the surface. Develop a model for the growth of this ferrite layer. Your model should result in an equation for the rate of growth of the layer. Estimate how long it will take until the layer thickness exceeds 50 μm in thickness. Assume that in pearlite the thickness of the cementite plates is about 15% of the total thickness of the pearlite.

6.15   In my research group we have been making nickel sheets by sintering powders. One way of doing this involves making a tape of the powder particles embedded in a polymeric binder. By heating the tapes at 500 °C the polymer is burnt out. However, this leaves a coating of carbon on each nickel particle. The nickel particles have an average diameter of 50 μm while the average thickness of the coating is 30 nm.

(a) What is the average weight fraction of carbon in this system?

(b) Set up the boundary-value problem you would need to solve to understand the diffusion of carbon into the nickel particles, *i.e.* what is the appropriate geometry, and the initial and boundary conditions?

(c) Estimate how long it will take to absorb all of the carbon into the nickel particles at 500 °C. What will be the carbon concentration in the centre of the particles at this time?

(d) One way to remove the carbon is to anneal the particles in a C-free atmosphere at 800 °C. How long will it take until the C concentration at the centre of the particles drops below 0.01 wt%?

Data: $\rho_{Ni} = 8.9 \times 10^3 \, kg/m^3$; $\rho_C = 2.3 \times 10^3 \, kg/m^3$.

6.16   A Ni–Al alloy of eutectic composition (13.3 wt% Al) is directionally solidified by pulling it slowly out of a furnace. This results in a rod-like morphology, in which cylindrical rods of the γ'-phase (Ni_3Al) are embedded in a nickel-rich γ-matrix. The rod axes are all aligned with the direction of the solidification. Assume that they have a diameter of 2 μm.

(a) Determine the average spacing of the rods. Assume that the compositions and distributions are those given by the phase diagram at the melting temperature, 1387 °C, and that the densities of the γ- and γ'-phases are approximately equal.

(b) You now wish to increase the amount of γ'-phase by annealing at 1000 °C. Set up the boundary-value problem you would need to solve to determine how long it will take to reach a new equilibrium. Your answer should include:
  (i) a conceptual picture or a description of the geometry involved;
  (ii) a statement of the initial condition;
  (iii) a statement of the boundary conditions;
  (iv) a set of schematic curves indicating how the composition of the alloy will change with time and position.

(c) Assume that diffusion in the γ-matrix is much faster than in the γ'-phase. Discuss how this will affect the overall process. You may wish to modify the detailed shape of the schematic curves in light of this knowledge.

(d) Determine the diameter of the rods, once a new equilibrium is reached.

(e) Determine how long it will take to establish a new equilibrium.

Data: Assume that the relevant diffusion coefficients are given by:
$D_0 = 1.6 \times 10^{-4} \, m^2/s$ and $Q = 316 \, kJ/mol$ in the γ'-phase;
$D_0 = 0.16 \, m^2/s$ and $Q = 316 \, kJ/mol$ in the γ-phase.

6.17   A local steel manufacturer makes galvanized steel sheet for auto-
       motive body panels. These steels must have very low interstitial (*i.e.*
       carbon and nitrogen) content. As part of the processing the sheets are
       given a short anneal after galvanizing. This annealing process (called
       'galvanneal') also helps to reduce the dissolved interstitial content.
       Measurements suggest that the dissolved carbon concentration in
       the α-Fe is about 0.0019 wt% as the steel enters the galvanneal fur-
       nace. The galvanneal furnace is at 480 °C and the residence time of the
       steel is 37 s. The equilibrium concentration of carbon in the Al-killed
       steel used is given approximately by

$$\text{wt\% C} = 0.72 \exp\left(-\frac{Q_C}{RT}\right)$$

       where Q_C, the activation energy for carbon solubility, is equal to
       40.6 kJ/mol. The excess carbon precipitates as carbides on grain
       boundaries. A typical grain diameter in steels of this class is 12 μm.
       (a) Estimate how long it will take during the galvanneal process
            for the average carbon concentration in the α-Fe to approach
            equilibrium. Compare this with the time of process. Is the time
            of the galvanneal adequate to lower the carbon level to equili-
            brium?
       (b) Following the galvanneal, the steel is cooled to room temperature.
            This cooling process is equivalent to spending an additional 30 s at
            280 °C. Estimate the equilibrium carbon concentration at this
            temperature. What will be the average carbon concentration in
            the steel following cool-down? What will the carbon concentra-
            tion be in the centre of the grains.
6.18   (a) The growth rate of precipitates growing by solute rejection into a
            matrix phase slows down with time. The growth rate of a eutec-
            toid mixture such as pearlite, growing by cooperative growth, is
            constant with time. Explain why the time dependences of these
            two processes are different. Be as precise as possible.
       (b) Choose either one of these processes and develop a model for the
            growth rate of the precipitate. Be sure to state any assumptions
            that you make. You may take Fick's First and Second Laws as
            given, but derive all other equations that you need.
       (c) Consider the Zn–Al phase diagram.
            (i) Consider an alloy with 78 wt% Zn, equilibrated at 400 °C. It is
                 now cooled to 250 °C. Assume that the separation between
                 plates as the new phase grows is 0.2 μm. How fast does the
                 interface move?

(ii) Consider an alloy with 30 wt% Zn, equilibrated at 400 °C. It is now cooled to 250 °C. How fast is the new phase growing 5 minutes after growth commences?

6.19   A cylindrical bar of iron, 2 cm in diameter, is suspended in a bath of manganese at 1350 °C.

(a) Describe the initial and boundary conditions relevant to this problem.

(b) Estimate how long it will take until the Mn concentration at the centre of the bar rises to 5 wt%.

(c) What will the average Mn concentration in the bar be at this time?

6.20   This problem is aimed at comparing the relative growth kinetics of spherical and plate-like precipitates growing by solute rejection. Start by reading **Example 6.4.** Extend this example by determining the relative growth time for a spherical precipitate and a plate-like precipitate of equal volume. In order to do this, you will need to make some assumption about the way in which the plates grow. For example, they might nucleate with a certain width, and maintain that width during growth, or they might maintain a constant aspect ratio (ratio of plate width to thickness) during growth. In either case you will need to introduce a separate parameter in your model. We will assume here that the plates maintain a fixed width (say d) and simply thicken during growth. Assume that the plates are square. Develop a general expression similar to the final expression of **Example 6.4.** Using the same parameters as in the **Example** and assuming a plate width of 25 nm determine the relative time for particle growth from an original volume of 300 nm^3 to 3000 nm^3. Comment on the significance of your answer in comparison with the result developed in the **Example.**

6.21   Stackpole Powder Metal Products make Fe–10 wt% Cr alloys by powder pressing and sintering. They start with separate powders of Fe and Cr (average particle diameter 30 μm) which are mixed and then pressed together. Following this stage, the material is almost fully dense. The material is then sintered at 1250 °C for 1 hour, in order to allow interdiffusion of Fe and Cr.

(a) Draw a schematic diagram illustrating the initial configuration at the start of the sintering process, and determine the average separation between Cr particles. You can assume that the density of iron and chromium are equal.

(b) The rate of diffusion is much faster in the α-phase than in the γ-phase. Estimate how long it will take until the α-particles come to equilibrium. (Use a reasonable degree of completion, say 95%, to determine this.) What will the size of the α-particles be at this time? (*Hint:* Cr is a substitutional solute in Fe. Moreover, it has almost the same size and atomic mass. Thus, you can neglect density differences.)

(c) Since the time calculated in part (b) is short, it is reasonable to assume that no diffusion has occurred in the γ-phase to this point. However, diffusion of Cr in the γ-phase will occur during the remainder of the 1 hour sintering cycle. Determine the concentration in the γ-phase midway between α-particles after 1 hour. Draw a schematic diagram of the concentration profile. (*Note:* If you were unable to obtain a reasonable answer for part (b), use a time of 5 minutes (this is not the correct answer).)

6.22 Sketch the region of the iron–carbon diagram that includes the eutectoid, then:

(a) Draw corresponding free energy vs composition plots for the α-, γ- and Fe_3C phases for temperatures just above, at, and just below the euctectoid temperature.

(b) Show the metastable phase boundaries that govern the growth of the α/Fe_3C eutectoid product (pearlite), and the composition difference that corresponds to the driving force for diffusion of carbon during pearlite growth. For a growth temperature of 675 °C, and a spacing S_α for the ferrite phase of 10 μm, estimate the velocity of the pearlite growth front.

(c) If the pearlite formed at 675 °C is quenched to 300 °C, calculate the time required to bring the α-phase to its new equilibrium carbon concentration. State any assumptions that you have made in arriving at your answer.

(d) It is possible to grow 'aligned' pearlite by passing a specimen of steel through a hot zone with a precisely defined temperature profile. A bar, treated this way to yield a growth interface temperature of 675 °C, is given to you, and you are asked to measure the self-diffusion of carbon (in the two-phase mixture) along the growth axis. Design an experiment to determine this quantity, and indicate how you would analyze the experimental results.

6.23 (a) The equilibrium concentration of foreign interstitials is described by the equation:

$$X_i^o = \frac{1}{1 + A \exp\left(\dfrac{\Delta G_i}{kT}\right)}$$

where X_i^o is the fraction of interstitial sites filled and A is a constant.

(i) Write down an expression for ΔG_i in terms of its components and define these precisely.

(ii) Derive an approximate version of the equation, valid at small concentrations.

(iii) Under what circumstances will the solubility X_i^o decrease with increasing temperature?

(b) A zirconium alloy containing 10 at.% H has been in service in a reactor at 300 °C for a long time. The reactor is then cooled to room temperature. Assuming that the alloy contains some coarse δ-ZrH$_2$ precipitates with an average spacing of 1 mm, set up a model to describe what you expect will happen at room temperature. Be as precise as possible (*i.e.* write down the initial conditions and boundary conditions, and draw a set of schematic diagrams).

(c) Determine (*i.e.* find a numerical value for) how long it will take for the hydrogen concentration everywhere in the alloy to drop to within 10% of its final value.

6.24   Suppose that a pure metal has been annealed at a temperature T_1 for sufficient time to establish an equilibrium vacancy concentration, $X_v^o = X_{v1}$. The temperature is now lowered to T_2, at which temperature the equilibrium vacancy concentration is X_{v2}. Assume that the vacancy concentration at T_2 changes with time according to a relationship of the form

$$X_v = a + b\exp(-t/\tau).$$

Determine an expression for the diffusion coefficient of atoms as a function of time at T_2. Be as precise as possible and define all terms used. Your final equation should not contain a and b, but their values in this case (*i.e.* you need to first determine what form a and b will take on in this problem).

Part C

Mass transport in concentrated alloys and fluids

Up to now we have only considered diffusion in solids. Even then the approach we have developed is strictly valid only for dilute alloys. Under these conditions diffusion is dominated by the solute alone. As noted in the introductory chapter the situation is a bit more complicated for a concentrated alloy, for which the solute diffusivity must be replaced by something called the 'interdiffusion coefficient'. In this part we will return to this concept and assess the basis for it. This will lead us to consider what happens when large concentration differences exist across a phase, and how we can accommodate a substantial change of the diffusion coefficient by utilizing the concept of counter diffusion.

We will then turn our attention to mass transport in fluids. Counter diffusion also occurs in fluids. However, additional effects occur because of the possibility of convective flow. This may involve either natural convection, driven by temperature and/or density differences within the fluid, or forced convection in which the fluid is made to flow at a fixed velocity (*e.g.* by a pump).

One final 'wrinkle' we will incorporate in this part involves the possibility of internal reactions. In the problems we have considered so far reactions have occurred only at well-defined interfaces. But in a fluid especially reactions can occur in a more diffuse way. Since this affects the concentration of each species locally it will clearly influence the driving force for diffusion.

As a result of incorporating these additional considerations we will arrive at a more general constitutive equation for mass transport, which will contain Fick's Laws as a special case.

Chapter 7

Diffusion in concentrated alloys and fluids

As knowledge grows, wonder deepens.
Charles Morgan

How much finer things are in composition than alone.
Ralph Waldo Emerson, *Journals*

In this chapter we address a range of issues related to mass transport when counter diffusion is possible. This can occur in solid alloys containing a relatively large concentration of a substitutional solute, such that solute diffusion requires a significant compensating diffusion of the solvent. It also occurs in fluids in which both counter diffusion and convective flow can occur. In this chapter we will consider only quiescent liquids (i.e. those in which convection due to external forces is absent). We will see that counter diffusion and convection are in fact similar in their impact on mass transport. This will enable us to develop a general framework in which it is possible to treat a wide range of problems. One process we will encounter in this chapter involves the use of diffusion couples, in which two materials are placed in contact such that interdiffusion occurs. Diffusion couples represent the second most common process for microstructural manipulation in the solid state (the first being heat treatment, as discussed in Chapter 6). We will also consider a range of problems in which a species diffuses into a binary mixture and reacts with one of the elements in this mixture at a well-defined front. Examples include evaporation and internal oxidation.

7.1   **Concept of counter diffusion**

Up to now we have been concerned primarily with diffusion in relatively dilute alloys. Under such conditions we could ignore the interrelationship between the diffusion coefficients of the various species that make up an alloy. But this is valid only in dilute alloys wherein diffusion is dominated by the diffusivity of the solute. In the case of an interstitial species, such as C in Fe, the spacing between carbon atoms is always quite large (compared to the lattice spacing). Therefore, the carbon atoms are able to move around quite independently of one another, and of the iron atoms.

In substitutional alloys, however, the diffusion of a species in one direction must be accommodated by the diffusion of another species in the opposite direction. Only in this way can the number of lattice sites be conserved, both locally and globally. Quantitatively, we can write for an alloy containing $\mathscr{N}$ different species (plus vacancies)

$$C = \sum_{i=1}^{\mathscr{N}} C_i + C_v \approx \sum_{i=1}^{\mathscr{N}} C_i, \tag{7.1}$$

where C_i is the concentration (in moles per unit volume) of each species, C_v that of the vacancies and C is the total concentration.[¶] This relationship couples the diffusion flux for each species to that of the other species in the system. For example, in a simple A–B binary alloy (and assuming a uniform vacancy concentration)

$$J_A = -D_A \frac{\partial C_A}{\partial y}, \tag{7.2}$$

while

$$J_B = -D_B \frac{\partial C_B}{\partial y}$$

$$= +D_B \frac{\partial C_A}{\partial y}. \tag{7.3}$$

The second part of eq. (7.3) is due to the fact that the total concentration is constant, *i.e.* $C_A + C_B = C$. In other words, the flux of A atoms must be correlated with a *counter diffusion* of B atoms.

In most instances, the diffusivity of A and B will be different. Thus, eqs. (7.2) and (7.3) indicate that the fluxes of these two species are in exact proportion to the diffusivities, *i.e.*

$$\frac{J_A}{J_B} = -\frac{D_A}{D_B}. \tag{7.4}$$

¶ Normally, the vacancy concentration is sufficiently small that we can neglect it in such molar balances, and we will generally do so here.

In other words, there will be a net flux, equal to $(J_A - J_B)$ in the solid. This will cause the specimen as a whole to *drift*, *i.e.* it will move with respect to an external frame of reference (y-axis on Fig. 7.1).

A similar result must occur in fluids. When one species flows within the fluid, a counter flow of other species must result, in order to maintain a uniform density (in the case of a liquid) or a uniform pressure (in the case of a gas). Since, in general, the various diffusivities are different, this will lead to a net motion in the fluid.

In addition, fluids can undergo *convection*. This differs from diffusion (the random motion of molecules within a body) in that convection involves the co-ordinated motion of a large number of molecules. Drift caused by counter diffusion is a form of convection. However, fluids can also undergo convection due to external force (*e.g.* temperature gradients or mechanical pumping). Whatever the cause, both drift and convection result in a body exhibiting an overall velocity.

This is illustrated in Fig. 7.1. In Fig. 7.1(a) a couple has been made from two materials A and B. A set of inert markers is placed between the original pieces in order to define the interface. This enables us to distinguish between an external (y-axis) frame of reference and an internal (η-axis) frame of reference. In this case we assume that $D_A \gg D_B$, for which there is a net drift to the right of velocity v. In Fig. 7.1(b), diffusion occurs in a fluid flowing with velocity v in a pipe. This case is phenomenologically similar to the previous

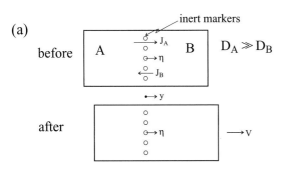

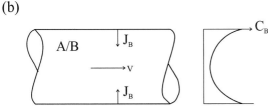

Figure 7.1 (a) A diffusion couple made from two soluble species A and B exhibits drift at a velocity v, with respect to the inert markers placed at the interface. (b) Diffusion in a fluid moves through a pipe at velocity v.

one, and as we will see it is possible to use the same conceptual framework to study both problems. Note, however, that in this case the direction of convective flow is different from that for the diffusive flux, as is often the case. Since both convection and diffusion are represented by vector quantities there is no conceptual difficulty here. But we must remain aware of this possibility.

In either of these two problems, we can separate the total flux of a species i, as seen from outside the body, into two terms – one concerned with diffusion, the other with the rigid motion of the body (whether by drift or convection). We determined back in Section 1.6 in fact that a steady drift at a velocity v produces a flux equal to $C_i \cdot v$. Since we derived this result based solely on the concept of conservation of matter, it is equally applicable in this context. If we add this to the diffusive flux the total is

$$N_i = J_i + C_i v, \tag{7.5}$$

where

$$J_i = -D_i \frac{\partial C_i}{\partial y}. \tag{7.6}$$

Here we have introduced a new parameter N, which we will use to distinguish the total flux in the presence of drift, from the diffusion flux alone, for which we will continue to use J. Note that we can always relate the flux of any species N_i to the average velocity of that species, by the same conservation principle we have used previously, so that

$$N_i = C_i v_i. \tag{7.7}$$

Moreover, if we sum eq. (7.5) for all species we can relate the velocity of each species to the total velocity. By definition, the sum of the diffusive fluxes (*i.e.* the J_i terms in eq. (7.5)) must be zero; otherwise they would contribute to drift. Thus,

$$N = Cv = \sum_i C_i v_i, \tag{7.8}$$

where the summation is over all species i in the material. In fact, this expression allows us to precisely define what we mean by overall drift velocity, as

$$v = \frac{\sum_i C_i v_i}{C} = \sum_i X_i v_i. \tag{7.9}$$

Now, throughout this development we have used molar concentrations, X_i. Thus v represents the average molar velocity. We could just as easily have developed these relationships in terms of mass concentrations, ρ_i. The resulting equations are:

$$n_i = j_i + \rho_i v^* = \rho_i v_i, \tag{7.10}$$

with

$$j_i = -D_i \frac{\partial \rho_i}{\partial y}; \tag{7.11}$$

and

$$n = \rho v^* = \sum_i \rho_i v_i, \tag{7.12}$$

where

$$v^* = \frac{\sum_i \rho_i v_i}{\rho} = \sum_i X_i^* v_i \tag{7.13}$$

is the average mass velocity, also called the 'physical velocity', and we have used n_i and n to denote total flux in units of mass per unit area per unit time. X_i^* is of course the weight fraction of species i.

It is also useful to combine eqs. (7.5) and (7.9) to obtain a relationship that directly connects the total and diffusive fluxes. Thus,

$$J_i = N_i - C_i \frac{\sum_k C_k v_k}{C} = N_i - X_i \sum_k N_k \tag{7.14}$$

on a molar basis, while on a mass basis

$$j_i = n_i - X_i^* \sum_k n_k. \tag{7.15}$$

It is worth noting that for dilute solutions, these equations reduce to

$$J_i = N_i, \quad X_i \ll 1$$

and

$$j_i = n_i, \quad X_i^* \ll 1. \tag{7.16}$$

These equations provide a general framework for solving diffusion problems in the presence of drift. We are now ready to look at some specific cases.

7.2   Kirkendall effect in solids

7.2.1   Interdiffusion coefficient

We start by considering the problem of diffusion in a binary solid. To proceed, we will consider a diffusion couple, in which two materials are placed in metallurgical contact with one another and allowed to interdiffuse. We will assume that the two materials exhibit complete solid solubility in one another at the temperature of interest. This happens, for example, if we join Cu and Ni, or MgO and NiO. Alternatively, we might couple two alloys of different compositions but within the same phase field. An example of this would be pure Cu with Cu–30 wt% Zn, both of which are in the α-brass field. When systems such as these are joined, there is no phase equilibrium at the interface. The concentration therefore changes continuously across the initial boundary and the actual interface is lost, or at least becomes hard to define and is difficult to see metallographically. We are nevertheless interested in the rate of the interdiffusion which occurs. If the range of compositional change across the couple is fairly large, we cannot neglect changes in the diffusion coefficient and the problem becomes more complex. In the Cu–Ni system for example, at 1000 °C, the self-diffusion coefficient of copper is about 100 times that of pure nickel. Moreover, the diffusion coefficient of copper within Cu–Ni alloys changes by a factor of 20 on going from pure Cu to pure Ni. This is too large a variation to ignore.

Before discussing this in detail, let us think about the consequences of having a difference in the diffusion coefficients of the two species. In the case we have been discussing, once we put copper and nickel in contact, Ni atoms will diffuse into the Cu-rich region and vice-versa. However, the rate of diffusion of Cu will be much faster. Therefore there will be a net migration towards the Ni-rich end of the specimen.

One way of envisaging the effect of this involves the use of markers. Suppose that we were to place a number of fine wires between the two blocks of A and B as we are making the diffusion couple. These wires would have to be inert, *i.e.* they would be made out of a material which did not react with or dissolve in A and B. That way, the wires simply serve to mark the position of the interface. The flux across this interface is imbalanced. More atoms are flowing towards the B side. If we were to suspend the couple from one of the wires, it would actually move towards the B-rich end (see Fig. 7.1(a)). This phenomenon is called the Kirkendall effect. The bulk movement of the specimen constitutes a process of *drift*. The questions we need to address are, first, how do we separate diffusion from drift and, second, is there a single diffusion coefficient which characterizes this behaviour?

The total flux of species A (see eq. (7.5)) is equal to

$$N_A = -D_A \frac{\partial C_A}{\partial y} + vC_A. \qquad (7.17)$$

Now when we derived Fick's Second Law, we showed that

$$\frac{\partial C}{\partial t} = -\frac{\partial J}{\partial y}.$$

This equation arises from a simple statement of conservation of matter by considering the total flux of solute through an elemental region. In the presence of drift this relation is still valid. However, we need to replace the diffusive flux J by the total flux N, i.e.

$$\frac{\partial C}{\partial t} = -\frac{\partial N}{\partial y}.$$
(7.18)

This represents a general result. The form which we developed previously is just a special case of this for dilute alloys. Thus we can substitute this into eq. (7.17) to get

$$\frac{\partial C_A}{\partial t} = \frac{\partial}{\partial y}\left[D_A \frac{\partial C_A}{\partial y} - vC_A\right].$$
(7.19)

If we do the same thing for component B, then:

$$\frac{\partial C_B}{\partial t} = \frac{\partial}{\partial y}\left[D_B \frac{\partial C_B}{\partial y} - vC_B\right].$$
(7.20)

But remember that the total concentration $C_A + C_B = C$, is a constant. Therefore,

$$\frac{\partial C_A}{\partial t} + \frac{\partial C_B}{\partial t} = 0.$$
(7.21)

If we therefore substitute eqs. (7.19) and (7.20) into this, we can solve for the drift velocity:

$$v = \frac{1}{C}(D_A - D_B)\frac{\partial C_A}{\partial y}.$$
(7.22)

i.e. the drift velocity is directly related to the difference between the diffusion coefficients.

We can now substitute this back into eq. (7.19) to get

$$\frac{\partial C_A}{\partial t} = \frac{\partial}{\partial y}\left[D_A \frac{\partial C_A}{\partial y} - D_A \frac{C_A}{C}\frac{\partial C_A}{\partial y} + D_B \frac{C_A}{C}\frac{\partial C_A}{\partial y}\right]$$

$$= \frac{\partial}{\partial y}\left[(D_A X_B + D_B X_A)\frac{\partial C_A}{\partial y}\right],$$
(7.23)

where X_A is the mole fraction of A, and X_B is the mole fraction of B, $X_B = 1 - X_A$. If we make the substitution

$$\tilde{D} = X_B D_A + X_A D_B$$
(7.24)

then the result has the form of Fick's Second Law

$$\frac{\partial C_A}{\partial t} = \frac{\partial}{\partial y}\left(\tilde{D}\frac{\partial C_A}{\partial y}\right).$$
(7.25)

However, the diffusion coefficient of component A has been replaced by a new parameter $\tilde{D}$ called the *interdiffusion coefficient*. This is a weighted average of

D_A and D_B. It is weighted, however, by the concentration of the opposite species. This means that the interdiffusion coefficient is dominated by the diffusion coefficient of the dilute component. Thus, our previous assumption of using only the diffusion coefficient of the solute is valid for dilute solutions. Moreover, this makes sense once you think about it. Consider a solution with 1 mole% A in B. The A atoms are surrounded by B atoms, and almost all of the jumps that are made by these atoms result in the diffusion of A. B atoms, on the other hand, are mostly surrounded by other B atoms. They have fewer opportunities to make jumps that lead to new configurations. They therefore have a much smaller effect on the diffusion kinetics.

An example of how the interdiffusion coefficient changes with composition is given in Fig. 7.2, for the case of copper and nickel. The interdiffusion coefficient has been calculated from the experimentally measured values for copper and nickel diffusion in various alloys using eq. (7.24).

Example 7.1: Interdiffusion coefficients

In a Ni–alloy containing 10 at.% Cu, estimate by how much the diffusion coefficient at 1000 °C differs from that for the dilute limit.

In the dilute limit the diffusion coefficient in the alloy is given by that for Cu in Ni. At 1000 °C, this is about $3 \times 10^{-14}\,\mathrm{m}^2/\mathrm{s}$ (from Fig. 7.2). For a 10 at.% alloy the diffusion coefficient of Cu is about $8 \times 10^{-14}\,\mathrm{m}^2/\mathrm{s}$. From Fig. 7.2 we see that the diffusion coefficient of Ni in the alloy is about 1/20th that of Cu for any alloy composition. Therefore, in the alloy the interdiffusion coefficient is

$$\tilde{D} = X_{Cu}D_{Ni} + X_{Ni}D_{Cu}$$

$$= X_{Cu}D_{Cu}/20 + (1 - X_{Cu})D_{Cu}$$

$$= D_{Cu}[1 - X_{Cu} + (X_{Cu}/20)]$$

$$= 0.895 D_{Cu} = 7.2 \times 10^{-14}\,\mathrm{m}^2/\mathrm{s}.$$

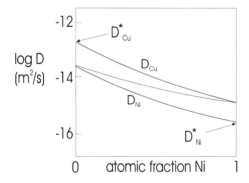

Figure 7.2 The diffusion coefficients (solid lines) for Cu and Ni diffusion in Cu–Ni alloys along with the interdiffusion coefficient (dotted line), plotted as a function of composition. D_{Cu}^* and D_{Ni}^* are the self-diffusion coefficients measured in pure Cu and Ni, respectively, using tracer diffusion measurements.

From this exercise we learn that there is a small decrease (about 10%) in the diffusivity due to the reduced concentration of Cu in the alloy. However, this is more than compensated by the increased mobility of the Cu atoms as the Cu concentration increases.

7.2.2 Diffusion couples

We will now consider how to apply this new diffusion coefficient. The simplest case involves problems in which the interdiffusion coefficient is independent of concentration within a given phase. Such may be the case, for example, when diffusion occurs across an intermetallic phase which exists only over a narrow range of stoichiometry (*e.g.* the $CuAl_2$ phase exhibits a range of only a few mole percent). In this case the result is rather trivial. Since $\tilde{D}$ is now a constant it can come outside the parentheses in eq. (7.25). Therefore, we simply need to use directly the solution for diffusion couples that we developed in Sections 3.3 and 3.6.

A more interesting situation occurs when the variation in $\tilde{D}$ is too great to be ignored, as would be the case if, for example, we joined pieces of pure nickel and pure copper together. If we know $\tilde{D}$ fully as a function of composition, then the diffusion equation, eq. (7.25), can be solved. However, the solution is likely to be complex (depending on the mathematical form of the diffusivity–concentration relationship). Thus one may need to resort to numerical techniques (such as finite-difference or finite-element methods) to obtain a valid solution. While this is numerically challenging, it does not present any new conceptual challenges and we will not go further along this line here.

It is interesting, however, to consider the opposite problem of how one determines the $\tilde{D}$–C relationship, from measured concentration profiles. The process of analyzing such data is called the *Matano–Boltzmann* method. We first transform eq. (7.25) into an ordinary differential equation by recognizing that the solution always depends on position and time in the form of $y/\sqrt{t}$. We therefore denote this by the symbol λ and substitute. The right-hand side of eq. (7.25) is transformed and the result is

$$\frac{\partial C}{\partial t} = -\frac{\lambda}{2t}\frac{\partial C}{\partial \lambda}, \tag{7.26}$$

while

$$\frac{\partial C}{\partial y} = \frac{1}{\sqrt{t}}\frac{\partial C}{\partial \lambda}. \tag{7.27}$$

Substituting these into eq. (7.25) gives

$$-\frac{\lambda}{2}\frac{dC}{d\lambda} = \frac{d}{d\lambda}\left(\tilde{D}\frac{dC}{d\lambda}\right). \tag{7.28}$$

We now revisit the diffusion couple pictured in Fig. 3.11, in which two semi-infinite blocks of materials of differing composition are joined together. Since the starting pieces do not need to be pure materials (so long as they are mutually soluble), we denote the initial concentrations as C_1 and C_2, respectively. Thus

$$\text{I.C.} \quad C = C_1, \qquad y < 0$$
$$C = C_2, \qquad y > 0.$$

Similarly, the boundary conditions are

$$\text{B.C.} \quad C = C_1, \qquad y \to -\infty$$
$$C = C_2, \qquad y \to \infty.$$

Equation (7.28) can simply be integrated from one side to any point in the interdiffusion region.

$$-\frac{1}{2} \int_{C_1}^{C} \lambda \, dC = \tilde{D}\left(\frac{dC}{d\lambda}\right), \qquad (7.29)$$

where we have used the limit that the concentration gradient goes to zero as C approaches C_1. This result can be inverted to determine $\tilde{D}$. Moreover, it is clear from this result that

$$\int_{C_1}^{C_2} \lambda \, dC = 0. \qquad (7.30)$$

One of the difficulties with this type of couple lies in defining a frame of reference, since the original interface becomes diffuse, and cannot be seen metallographically. This analysis, however, allows us to define the *Matano* interface for which

$$\int_{-\infty}^{0} \lambda \, dC = \int_{0}^{\infty} \lambda \, dC. \qquad (7.31)$$

This is shown schematically in Fig. 7.3. The definition of the Matano interface requires that the hatched region on either side of the interface be of equal area. In other words, the amount of solute lost from one side of the Matano inter-

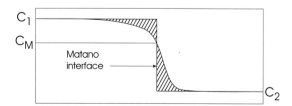

Figure 7.3 A schematic view of the diffusion profile across a couple with variable inter-diffusion coefficient, illustrating the position of the Matano interface (courtesy G. R. Purdy).

face is exactly equal to the solute absorbed on the other side. This therefore represents the original position of the interface. The concentration at the interface is given by C_M on the diagram and in general it will not be equal to $\frac{1}{2}(C_1 + C_2)$.

7.3 Solid-state diffusion couples involving immiscible phases

We now want to consider what happens when two materials which are mutually insoluble are placed into metallurgical contact at high temperature. Once interdiffusion starts to occur, a well-defined interface develops, possibly with one or more intermediate phases, as indicated by the relevant phase diagram. For example, when steel is dipped into liquid zinc, a series of intermetallic phases develop on the surface of the sheet. Alternatively, we may want to know what happens when MgO and Cr_2O_3 powders are mixed and then fired to produce a refractory brick.

7.3.1 Simple diffusion couples for immiscible solids with no intermediate phases

We begin with systems involving two materials that are not completely soluble in each other but which do not exhibit any intermediate phases. A representative phase diagram of this type is shown in Fig. 7.4. Let us consider two pure materials A and B which are placed in contact with a planar interface, as shown in Fig. 7.5.

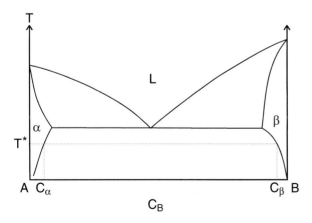

Figure 7.4 A schematic binary phase diagram. A diffusion couple is formed by placing pure A and pure B in contact at temperature T.

As shown in the phase diagram (Fig. 7.4), material A has some solubility in B (the β-phase) and B has some solubility in A (the α-phase). The most common systems of this type are those with simple eutectic phase diagrams, such as Pb–Sn.

We now ask what happens when A and B are placed together at a sufficiently high temperature (such as T^* in Fig. 7.4) for diffusion to occur? The initial conditions are clear. On one side of the interface (which we put at $y = 0$) we have pure A; on the other pure B, *i.e.*

$$\text{I.C.} \quad \begin{aligned} C &= C^{\mathrm{B}}, & y &< 0, & t &= 0 \\ C &= 0, & y &> 0, & t &= 0, \end{aligned}$$

where C^{B} represents the concentration (in moles per unit volume) of pure B. Here C represents the concentration of component B. There is nothing special about B in this case however. We could just as easily have chosen to define the problem in terms of component A.

In thinking about the boundary conditions for this problem we have to first realize that diffusion will now occur in both the α- and β-phases. Moreover, because these are different phases, the diffusion coefficient will be different in each. We therefore have two separate problems, each with its own set of boundary conditions. At short times, we can write the boundary conditions as:

$$\text{B.C.} \quad \begin{aligned} \text{in the } \alpha\text{-phase: } C &= 0, & y &\to +\infty \\ C &= C_\alpha, & y &= 0 \\ \text{in the } \beta\text{-phase: } C &= C^{\mathrm{B}}, & y &\to -\infty \\ C &= C_\beta, & y &= 0. \end{aligned}$$

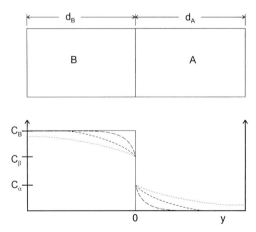

Figure 7.5 Illustration of a diffusion couple between two materials A and B. The phase diagram appropriate to this case is shown in the previous figure.

At long times, when the diffusion profile extends to the ends of the bar, we need to modify the boundary conditions. If the ends are impermeable, *i.e.* the material has a slow rate of evaporation, then the flux will be zero at these locations, and thus we have:

$$
\begin{aligned}
\text{B.C.} \quad &\text{in the } \alpha\text{-phase: } \partial C/\partial y = 0, && y = +d_A \\
&\qquad\qquad\qquad\; C = C_\alpha, && y = y_I \\
&\text{in the } \beta\text{-phase: } \partial C/\partial y = 0, && y = -d_B \\
&\qquad\qquad\qquad\; C = C_\beta, && y = y_I
\end{aligned}
$$

Here, d_A and d_B are the original widths of the plates A and B, respectively, as shown in Fig. 7.5. These conditions would also be applicable if a periodic structure of A and B plates were studied, except that d_A and d_B would then be replaced by the half-width of the A and B plates, respectively. Note that the position of the interface has now been set to be variable, $y = y_I$, in order to recognize that the interface may move during this process. We can determine the rate of interface motion by using the principle of mass conservation as we did earlier when considering precipitate growth by solute rejection. We again consider what happens when the interface moves by an amount dy_I over a time interval dt. The amount of B which moves into the α-phase is $J_\alpha \cdot A \cdot dt$. Similarly, the amount of B moving out of the β-phase is $J_\beta \cdot A \cdot dt$. Thus the total build-up of B at the interface is equal to $(J_\beta - J_\alpha) \cdot A \cdot dt$. This excess material must go towards creating more of the phase which is rich in B, *i.e.* the β-phase. As this happens the interface moves by dy_I. The amount of B which is taken up is therefore equal to $(C_\beta - C_\alpha) \cdot A \cdot dy_I$. By equating these two terms we can determine the rate of interface motion to be

$$
\frac{dy_I}{dt} = \frac{J_\beta - J_\alpha}{C_\beta - C_\alpha}. \tag{7.32}
$$

We see that if $J_\beta > J_\alpha$ there is a net build-up of solute at the interface. Thus the solute-rich (β) phase must grow at the expense of the solute-poor (α) phase. Therefore the interface moves to the right, as predicted by the equation. If $J_\beta < J_\alpha$, the opposite holds. We can compare this result with that which we derived for solute rejection (eq. (6.5)). The only difference is that we now have two fluxes at the interface. The result is otherwise identical. Thus, the case of precipitate growth by solute rejection is a special case of a moving interface in which diffusion occurs only in one of the phases.

The final issue to resolve is what diffusion coefficients we need to use. In general this will be the interdiffusion coefficient $\tilde{D}$ for each phase. How we apply this in practice, however, will depend on the exact nature of the problem. If A, for example, is a pure material and the solubility of B in the α-phase is not too large then, to a good approximation, $\tilde{D}_\alpha = (D_B)_\alpha$ the diffusion coefficient of B in the

α-phase, in the dilute limit. If the solubility of B in the α-phase is large we may want to calculate $\tilde{D}_\alpha$ more precisely. But since this may depend on composition, the analysis can become considerably more complex. Finally, if one of the starting phases is not a pure material but rather a solid solution, then we will need to use the interdiffusion coefficient for that phase.

7.3.2   Couples with intermediate phases

In many binary systems there exist one or more intermediate stable phases between the two pure components. We now want to consider what happens in this situation. We will start by looking at a generic phase diagram of this type, shown in Fig. 7.6. What will happen if we place a piece of A in contact with a piece of B at temperature T^*? Clearly B will diffuse into the A region, and vice-versa. But the phase diagram tells us that as this occurs a thin layer of the intermediate β-phase will form between the two terminal α- and γ-phases. As diffusion proceeds, this layer will thicken. The situation is somewhat reminiscent of that for the growth of surface films which we considered earlier. We called the growth of these films a 'pseudo' steady-state process, because the diffusion is rapid across thin regions and can keep up with the movement of the interface. The same thing occurs here. Thus we can use the pseudo-steady-state assumption, which greatly simplifies the analysis.

The initial conditions for this problem are the same as they were before, *i.e.*

$$\text{I.C.} \qquad \begin{aligned} C &= C^{\mathrm{B}}, & y &< 0, \quad t = 0 \\ C &= 0, & y &> 0, \quad t = 0. \end{aligned}$$

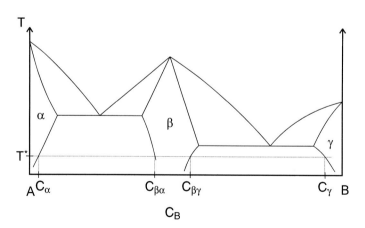

Figure 7.6 A schematic binary phase diagram of a system having an intermediate phase. In this case it consists of two simple eutectic reactions.

We now have *three* sets of boundary conditions, one for each phase. For short times these can be written as:

I.C.   in the γ-phase: $C = C^{\mathrm{B}}$, $y \to -\infty$

$C = C_{\gamma}$, $y = y_{\gamma}$

in the β-phase: $C = C_{\beta\gamma}$, $y = y_{\gamma}$

$C = C_{\beta\alpha}$, $y = y_{\alpha}$

in the α-phase: $C = C_{\alpha}$, $y = y_{\alpha}$

$C = 0$, $y \to \infty$.

The diffusion profile which results from this is shown in Fig. 7.7. Of course, if we wish to consider solutions at long times, we can replace the boundary conditions at the end of the bar by zero flux conditions, as we did before. However, the question we usually want to address in situations such as this is: how fast does the intermediate layer grow? Furthermore we are generally interested in the early stages of this process.

Now, to proceed further, we need to think in more detail about diffusion processes in intermetallic or compound phases (*e.g.* $CuAl_2$, $MgAl_2O_4$, etc.). It is important to know whether the elements in the phase form a random solid solution or develop an ordered structure with each element assigned to a specific lattice position. Generally non-metallic compounds are highly ordered.

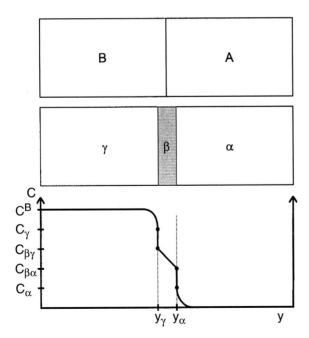

Figure 7.7 The diffusion profile for a diffusion couple containing an intermediate phase β. The thickness of the interfacial layer of β is $y_{\alpha} - y_{\gamma}$.

Intermetallic compounds, however, may be ordered or not and the degree of order often varies with temperature.

Why is this important? Well, for an ordered compound, *e.g.* NiO, the diffusion fluxes of the two species are independent of one another, and we can write a separate conservation equation for each sublattice.

In the case of NiO, this would be

$$C = C_{\text{Ni}} + C_{\text{v}}^{\text{Ni}}$$

for the nickel sublattice (C_{v}^{Ni} being the vacancy concentration in that sublattice), while

$$C = C_{\text{O}} + C_{\text{v}}^{\text{O}}$$

on the oxygen sublattice. In solving diffusion problems, we treat diffusion on each sublattice separately. Thus when we considered the oxidation of nickel in Section 2.3.1, we noted that the scale could grow at the Ni/NiO interface and the Ni/oxygen interface simultaneously. Moreover, the overall rate is controlled by whichever species (Ni^{2+} or O^{2-} ions) can diffuse more rapidly through a NiO scale. For solid solutions on the other hand, the fluxes are coupled as we have discussed and the interdiffusion coefficient can be used.

We will start with the case of a solid solution compound. If we assume a pseudo-steady-state condition within the β-layer, then a linear concentration profile exists across the β-phase, and the flux of B is given by Fick's First Law:

$$J_{\beta} = -\tilde{D}\,\frac{C_{\beta\alpha} - C_{\beta\gamma}}{y_{\alpha} - y_{\gamma}} = +\tilde{D}\,\frac{C_{\beta\gamma} - C_{\beta\alpha}}{L},$$

where L is now the (time-dependent) thickness of the intermediate layer, equal to $(y_{\gamma} - y_{\alpha})$. The motion of each interface is governed by the general equation we derived in the previous section. Thus

$$\frac{\mathrm{d}y_{\alpha}}{\mathrm{d}t} = \frac{J_{\beta} - J_{\alpha}}{C_{\beta\alpha} - C_{\alpha}} \tag{7.33}$$

and

$$\frac{\mathrm{d}y_{\gamma}}{\mathrm{d}t} = \frac{J_{\gamma} - J_{\beta}}{C_{\gamma} - C_{\beta\gamma}}. \tag{7.34}$$

When the layer is very thin, J_{β} must be very large and it will dominate. The growth of the layer thickness is given by the difference between the motion of each interface. Thus,

$$\frac{\mathrm{d}L}{\mathrm{d}t} = \frac{\mathrm{d}y_{\alpha}}{\mathrm{d}t} - \frac{\mathrm{d}y_{\gamma}}{\mathrm{d}t} \approx \frac{J_{\beta}}{C_{\beta\alpha} - C_{\alpha}} + \frac{J_{\beta}}{C_{\gamma} - C_{\beta\gamma}}.$$

If we substitute for J_β in this equation, the result is

$$\frac{dL}{dt} = \frac{\tilde{D}}{L}\left(\frac{C_{\beta\gamma} - C_{\beta\alpha}}{C_{\beta\alpha} - C_\alpha} + \frac{C_{\beta\gamma} - C_{\beta\alpha}}{C_\gamma - C_{\beta\gamma}}\right). \tag{7.35}$$

This should be familiar to you by now from other problems involving a pseudo-steady-state, except that the concentration term is now more complicated. But since all these concentrations are constants, this equation is again easily integrated over time, leading to the usual parabolic dependence of layer thickness on time, namely,

$$L^2 = 2\tilde{D}\left(\frac{C_{\beta\gamma} - C_{\beta\alpha}}{C_{\beta\alpha} - C_\alpha} + \frac{C_{\beta\gamma} - C_{\beta\alpha}}{C_\gamma - C_{\beta\gamma}}\right)t. \tag{7.36}$$

The situation for ordered compounds is similar, but differs in some of the details. In this case we treat the diffusion of each species independently. Thus, the flux of B across the β-phase is given by

$$J_\beta^B = -D_B^\beta\,\frac{C_{\beta\gamma} - C_{\beta\alpha}}{L}, \tag{7.37}$$

while that for A is

$$J_\beta^A = -D_A^\beta\,\frac{C_{\beta\gamma} - C_{\beta\alpha}}{L}. \tag{7.38}$$

Equations (7.33) and (7.34) still hold since they are simply statements of matter conservation, and once again we can assume that the process is dominated by flux across the β-phase. Thus the final result is

$$L^2 = 2\left(D_B^\beta\,\frac{C_{\beta\gamma} - C_{\beta\alpha}}{C_{\beta\alpha} - C_\alpha} + D_A^\beta\,\frac{C_{\beta\gamma} - C_{\beta\alpha}}{C_\gamma - C_{\beta\gamma}}\right)t. \tag{7.39}$$

Recall that when we discussed the growth of oxide layers on a surface, we noted that the rate of film growth will be controlled by the faster diffusing species. This will be the case here also.

Let us look now at some examples of this kind of problem.

Spinels. A number of binary oxide systems contain a single intermediate phase called a spinel. Two examples include the MgO–Al_2O_3 system and the MgO–Fe_2O_3 system.[†] If powders of MgO and Al_2O_3 are mixed together and fired at say $1700\,^\circ C$ then the spinel phase $MgAl_2O_4$ will form wherever the different powder particles are in contact. If the average composition of the mixture is in a range between about 70 and $80\,wt\%\,Al_2O_3$ then the body will eventually be converted entirely to spinel. If the composition lies outside this range then some of the primary powder will be left.

[†] The appropriate phase diagrams are contained in Appendix C.

Direct bonding in refractories. A similar situation is found in the MgO–Cr$_2$O$_3$ system. This system also forms a spinel phase. This is used to advantage to produce refractories for high-temperature furnaces. The intermediate spinel phase (MgCr$_2$O$_4$) forms between the MgO and Cr$_2$O$_3$ particles, producing a strong bond between them.

Surface coating. The superconductor Nb$_3$Sn can be made by dipping Nb wire into a bath of Sn. A reaction takes place at the surface in which Nb$_3$Sn is formed. The depth of the layer will depend on the temperature of the Sn bath, the residence time of the wire in the bath and the rate of diffusion through the Nb$_3$Sn layer. Similar situations are found in the Fe–Zn system, in which case Fe (or steel) is dipped into molten zinc which results in a series of intermetallic phases being deposited on the surface. This is the basis of hot-dip galvanizing. Another example involves the formation of mullite (Al$_2$O$_3$)$_3$ · (SiO$_2$)$_2$. Alumina particles can be coated with fine silica particles from an aqueous solution. When the body is then fired a conversion to mullite takes place.

In all of these problems, the intermediate phase is initially formed at points of contact between particles. In some cases, the reaction does not proceed to completion. In the manufacture of refractory bricks using MgO and Cr$_2$O$_3$-containing ores for example, the formation of a spinel layer at the interface is used to produce a strong bond between particles. Such materials are referred to as direct-bond refractories. If we know the diffusion coefficient within the intermediate phase, we can use eq. (7.36) or (7.39) to determine the time required to develop a bonding layer of a given thickness L.

If we are interested in seeing the reaction proceed to completion, the analysis is more complex. This is because eq. (7.36) applies strictly to the growth of a layer when the flux in the layer is much larger than in the parent phases. If we use this equation anyway, setting L equal to the average separation between the starting phase particles, we will underestimate the time required.

7.4   Diffusion in quiescent fluids

In a fluid, the atoms or molecules are no longer restricted to regular positions. Moreover, they are much more mobile. As a consequence, mass transport occurs not only by diffusion, but also by convection. In this section, however, we will consider transport in *quiescent* fluids, *i.e.* those for which convective flow is negligible.

7.4.1   Simple evaporation

Consider for example the evaporation of a gas A out of a column, as illustrated in Fig. 7.8. The column is designed to be tall and slender so as to limit convection.[¶] A carrier gas containing a known pressure of A mixed with an inert gas B, is passed over the top of the column.

Once this process begins, A will evaporate from the liquid, *i.e.* it diffuses up the column and leaves by way of the carrier gas. Moreover, B will diffuse down the column. After some time a steady-state will be established in which the concentrations of A and B become constant. The boundary conditions for this configuration are given by

$$
\begin{aligned}
\text{B.C.} \quad X_A &= X_{A1}, & y &= 0 \\
X_A &= X_{A2}, & y &= L.
\end{aligned}
$$

In this case, X_{A1} corresponds to the vapour pressure of the gas A in equilibrium with the pure liquid, at the temperature of interest. Moreover, the concentration X_{A2} is simply equal to the concentration of A in the carrier gas, provided that the flow rate is sufficiently rapid. If the flow is quiescent in the column, then the concentration of A will be a function of y only, and not of the radial position within the column. Moreover, mass conservation requires that the flux be a constant throughout the column, *i.e.*

$$\frac{\partial N_A}{\partial y} = 0. \tag{7.40}$$

We now use eq. (7.14) which, for a binary system, can be rearranged to give

$$N_i = J_i + X_i(N_A + N_B) \tag{7.41}$$

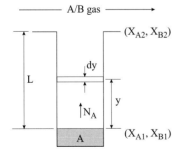

Figure 7.8 A slender evaporation column, of height L, contains a liquid of pure A. An A/B gas mixture is passed over the top of the column. B is non-reactive with A.

[¶] Experimental setups such as this are indeed used to measure the diffusion coefficient within gases and gas mixtures.

where the subscript i may be either A or B. Specifically, we can write the net flux of A as

$$N_A = -CD_A \frac{\partial X_A}{\partial y} + X_A(N_A + N_B),$$ (7.42)

where $C = C_A + C_B$, see eq. (7.1). In this particular system, once steady-state is established, $N_B = 0$. This is because the inert gas B can enter the column but has no place to exit.[¶] Thus, eq. (7.42) reduces to

$$N_A(1 - X_A) = -CD_A \frac{\partial X_A}{\partial y}.$$

Separating the flux and concentration components gives

$$\frac{N_A}{CD_A} = \frac{-1}{1 - X_A} \frac{\partial X_A}{\partial y}.$$ (7.43)

Since the two sides of this equation are independent, they must each equal a constant, say α. We can integrate the right-hand side, yielding

$$\ln(1 - X_A) = \alpha y + \beta$$

where β is a constant of integration. We now evaluate this expression subject to the boundary conditions to determine α and β. The result is

$$\ln\left(\frac{1 - X_A}{1 - X_{A1}}\right) = \frac{y}{L} \ln\left(\frac{1 - X_{A2}}{1 - X_{A1}}\right).$$ (7.44)

This expression is plotted in Fig. 7.9.

What we really want to know of course, is how fast A evaporates and leaves the column. This is given by the flux N_A. We refer back to eq. (7.43), from

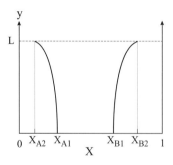

Figure 7.9 A plot of X_A and X_B vs height y in a column. The logarithmic profile is the result of counter diffusion in a stagnant column.

[¶] This may seem confusing at first. Since $\partial X_B/\partial y$ is not equal to zero within the column, there must be a diffusive flux of B towards the bottom. Since $N_B = 0$, this must be exactly compensated by a vertical drift of B molecules. Physically, this is caused by the motion of A out of the column, creating a net upward drift that carries B molecules along with it.

which the following integral can be derived:

$$\frac{N_A}{CD_A} \int_0^L dy = -\int_{X_{A1}}^{X_{A2}} \frac{dX_A}{1 - X_A}.$$ (7.45)

This leads to an expression of the form:

$$N_A = \frac{CD_A}{L} \ln\left(\frac{1 - X_{A2}}{1 - X_{A1}}\right) = \frac{CD_A}{L} \ln\left(\frac{X_{B2}}{X_{B1}}\right).$$ (7.46)

This expression looks rather like Fick's First Law, except that the concentration difference $(X_{B2} - X_{B1})$ has been replaced by the difference in the logarithmic concentration, $i.e.$ $\ln(X_{B2}/X_{B1}) = \ln X_{B2} - \ln X_{B1}$.

We can therefore rewrite the flux as

$$N_A = \frac{CD_A}{L} \frac{X_{B2} - X_{B1}}{(\Delta X_B)_{\ln}} = \frac{J_A}{(\Delta X_B)_{\ln}},$$ (7.47)

where

$$(\Delta X_B)_{\ln} \equiv \frac{X_{B2} - X_{B1}}{\ln\left(\dfrac{X_{B2}}{X_{B1}}\right)}$$ (7.48)

is called the *logarithmic mean driving force*. This represents the enhancement of the flux due to the counter flow of B into the column. Thus $D_A/(\Delta X_B)_{\ln}$ is analogous to the interdiffusion coefficient we developed previously. You should also note that in the dilute limit (*i.e.* when the vapour pressure of A is small compared with pressure of B in the carrier), $(\Delta X_B)_{\ln}$ approaches unity, and the Fick's First Law solution can be used directly.[¶] This is consistent with our earlier observations regarding the interdiffusion coefficient in dilute alloys. It also conforms to the limits developed earlier in this chapter (see eqs. (7.16)).

7.4.2 Evaporation involving reactions at a front

We now consider a slightly more complex case in which the evaporating species reacts with some component of the carrier gas. A good example is selenium, which is a liquid at room temperature.

Suppose that you wish to measure the evaporation rate of Se using the apparatus illustrated in Fig. 7.10. And let us further suppose that the carrier gas is simply air (*i.e.* a mixture of N_2 and O_2).[†] Now Se gas oxidizes readily to form SeO_2 which is also a gas:

$$Se_{(g)} + O_{2(g)} \rightarrow SeO_{2(g)}$$

[¶] You can easily convince yourself of this by recalling that $\ln x \approx 1 + x$ when x is close to 1.
[†] Actually any gas containing oxygen will do.

As the Se evaporates from the liquid at the base of the tube, it rises by diffusion until it meets the oxygen diffusing from the top. We can treat this process
using our current methodology only because the rate of reaction once Se and
O_2 meet is extremely rapid. Therefore as soon as Se comes in contact with
oxygen it oxidizes. This means that all of the reaction will occur at a single
height in the tube. Once a steady-state is established the height at which this
reaction occurs will be fixed at, say, $y = H$. We will also neglect the diffusion
of SeO_2 away from the point of reaction.

We are therefore left with two complete but distinct diffusion problems.
Below $y = H$, Se diffuses in a nitrogen carrier gas. This is governed by eq.
(7.40) subject to the boundary condition:

$$\text{B.C.} \qquad X_{Se} = X_{Se}^o, \qquad\qquad y = 0$$
$$X_{Se} = 0, \qquad\qquad y = H.$$

Here X_{Se}^o is the vapour pressure of Se above its liquid. Above $y = H$, O_2
diffuses in a nitrogen carrier gas, subject to a slightly different set of boundary
conditions

$$\text{B.C.} \qquad X_{O_2} = (X_{O_2})_L, \qquad\qquad y = L$$
$$X_{O_2} = 0, \qquad\qquad y = H.$$

Here, $(X_{O_2})_L$ is the concentration of oxygen in the carrier gas $((X_{O_2})_L = 0.21$ if
this is air). The solution to these problems follows directly from that given in
the previous section with the appropriate change of variable names. Thus the
rate of Se evaporation, from eq. (7.46), is

$$N_{Se} = -\frac{CD_{Se}}{H} \ln(1 - X_{Se}^o). \qquad (7.49)$$

However, the position of the reaction front H, is not known a priori. We

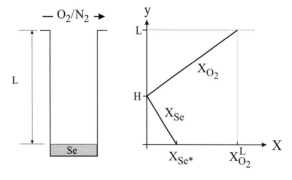

Figure 7.10 A narrow-tube evaporation column containing liquid selenium, over which
an O_2/N_2 mixture is passed. The O_2 and Se meet at height H where they
react rapidly.

therefore also need to calculate the oxygen flux

$$N_{O_2} = -\frac{CD_{O_2}}{L - H} \ln[1 - (X_{O_2})_L].$$ (7.50)

At $y = H$ these fluxes must be equal but of opposite sign. Thus,

$$H = \frac{L}{1 + \dfrac{D_{O_2}}{D_{Se}} \dfrac{\ln[1 - (X_{O_2})_L]}{\ln(1 - X_{Se}^o)}}.$$ (7.51)

We can now substitute this back into eq. (7.49) to get the rate of evaporation:

$$N_{Se} = -\frac{CD_{Se}}{L} \ln(1 - X_{Se}^o)\left[1 + \frac{D_{O_2}}{D_{Se}} \frac{\ln[1 - (X_{O_2})_L]}{\ln(1 - X_{Se}^o)}\right].$$ (7.52)

We can simplify this expression by noting that the equilibrium vapour pressure of Se is relatively small (*i.e.* $X_{Se}^o \ll 1$). Moreover, if we do the experiment, not in air, but with a low oxygen pressure such that $X_{O_2} \ll 1$, then

$$N_{Se} = \frac{CD_{Se}}{L} X_{Se}^o \left[1 + \frac{D_{O_2}(X_{O_2})_L}{D_{Se}X_{Se}^o}\right].$$ (7.53)

The term in large square brackets in both eqs. (7.52) and (7.53) represents the additional contribution due to reaction. This term is always greater than unity, which means that the reaction enhances the rate of evaporation, because the concentration gradient is steeper than without reaction.

7.4.3 Particle condensation during evaporation

Molten iron (or steel) sitting in a ladle or tundish will slowly evaporate. Even when a protective atmosphere such as argon is used, there is always sufficient oxygen present to oxidize the Fe vapour:

$$Fe_{(g)} + \tfrac{1}{2}O_{2(g)} \rightarrow FeO_{(s)}$$

Since the resultant oxide is solid, it condenses to form a 'fog' of FeO particles. Turkdogan *et al.*[¶] found experimentally that the rate of evaporation increases steadily as the oxygen partial pressure increases. Similar results were found not just for Fe but also Cu, Ni, Co, Mn and Cr.

This problem can be analyzed using the method just developed. The only change we need to make (besides replacing Se by Fe in all of the equations) is to note the different stoichiometry. This implies that at the interface $N_{Fe} = 2N_{O_2}$. To account for this, it is necessary to replace D_{O_2} by $2D_{O_2}$ in all instances where this occurs in eqs. (7.51)–(7.53). The result is otherwise unchanged.

¶ E. T. Turkdogan, P. Grieveson and L. S. Darken, *J. Metals,* **14**, 521 (1962); *J. Phys. Chem.,* **67**, 1647 (1963)

However, the experimental situation is not as clear. Since evaporation is not occurring in a column of known dimensions, the parameter L in eq. (7.52) is not known. Since the equilibrium vapour pressure and the diffusion coefficient for Fe vapour above pure molten Fe are known, we could use eq. (7.49), but only if the thickness of the Fe vapour layer were known. Turkdogan *et al.* estimate this by considering evaporation into a vacuum. In this case, the rate of evaporation is given by the Langmuir equation

$$N_i^{\max} = \frac{p_i^*}{\sqrt{2\pi RTm_i}},$$
(7.54)

where p_i^* and m_i are the vapour pressure and molecular weight, respectively, of the evaporating species. This sets an upper limit on the possible rate of evaporation, unhindered by diffusion away from the surface. This equation can be used along with eq. (7.49) to determine the minimum possible value of H. The result is

$$H^{\min} = \frac{CD_i}{p_i^*} \sqrt{2\pi RTm_i} \ln(1 - X_i^{\circ}).$$
(7.55)

However, this represents a limiting case. In the presence of a protective atmosphere, the diffusion of Fe vapour above the melt is hindered. As a result the flux away from the surface will be less than that given by eq. (7.54). This increases the thickness of the vapour layer H and the FeO fog will appear at a greater distance from the surface. Figure 7.11 shows some of the experimental work of Turkdogan *et al.* The flux of Fe from the surface has been measured by passing an argon/oxygen gas over the surface at fixed velocity

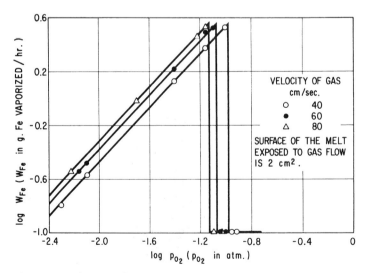

Figure 7.11 The rate of Fe evaporation from the surface of molten Fe at 1600°C is plotted as a function of the oxygen partial pressure for three gas flow rates (E. T. Turkdogan, P. Grieveson and L. S. Darken, *J. Metals,* **14**, 521, (1962)).

and collecting the FeO particles that are produced. The flux of Fe increases linearly with oxygen partial pressure as predicted by the Langmuir equation (eq. 7.54). It also increases slowly with increasing flow rate. Note, however, that above a certain pressure the evaporation drops to zero. This is because as the position for condensation H decreases the most favourable mechanism switches over to surface scale formation. Such scales are passivating as we have discussed earlier, and the rate of oxidation is greatly reduced.

Example 7.2: Thickness of FeO vapour layer above molten Fe

For an Fe melt at 1600 °C, determine the maximum rate of Fe evaporation and the corresponding thickness of the Fe vapour layer. Assume an argon/oxygen atmosphere with a total pressure of 1 atm.

Data: $p_{Fe}^* = 7.5 \times 10^{-5}$ atm.

$D_{Fe\,in\,Ar} = 5.6 \times 10^{-4}\,\mathrm{m^2/s}.$

From the Langmuir relation eq. (7.54), we can evaluate the maximum rate of evaporation. The vapour pressure is 7.5×10^{-5} atm $= 7.5$ Pa, while the molecular weight of Fe is 0.0558 kg/mol. Thus,

$$N_{Fe}^{max} = \frac{7.5}{\sqrt{2\pi \times 8.314 \times 1873 \times 0.0558}} = 0.102\,\mathrm{mol/m^2\,s}$$

This value is similar to the maximum evaporation rate of $0.088\,\mathrm{mol/m^2\,s}$ measured experimentally by Turkdogan *et al.*

To determine the corresponding minimum value for H, we need to know the total concentration in the gas phase C (mol/m^3). For an ideal gas $C = p/RT$ where $p = 1$ atm $= 1.01 \times 10^5$ Pa. The mole fraction of Fe at the surface is $X_{Fe}^o = p_{Fe}^*/p = 7.5 \times 10^{-5}$. Therefore, by substitution into eq. (7.49), we find that

$$H^{min} = -\frac{CD_{Fe}}{N_{Fe}^{max}}\ln(1 - X_{Fe}^o) \approx \frac{pD_{Fe}}{RTN_{Fe}^{max}}X_{Fe}^o$$

$$= \frac{(1.01 \times 10^5)(5.6 \times 10^{-4})(7.5 \times 10^{-5})}{8.314 \times 1873 \times 0.102}$$

$$= 2.67 \times 10^{-6}\,\mathrm{m} = 2.67\,\mathrm{\mu m}.$$

7.5   Near-surface internal precipitation in alloys

In Section 2.3.1 we considered the oxidation of a pure metal by the development of a continuous oxide scale on the surface. The oxidation of alloys can also proceed by such a path. However, other possibilities exist in this case. For example, following the oxidation of copper alloys containing aluminum, a fine dispersion of Al_2O_3 is found beneath the surface. Similar behaviour is found following the sulphidation of Fe–Mn alloys (now involving MnS particles). A related phenomenon involves the precipitation of vapour bubbles beneath the surface of alloys during annealing in certain gas atmospheres. For example, Ni–C alloys are found to contain CO_2 bubbles following oxidation, while Cu–O alloys develop water–vapour bubbles following high-temperature exposure to hydrogen.[¶] A common feature of these phenomena is that a uniform precipitate dispersion is found below the surface, which then ends abruptly at a well-defined depth. So, for the Cu–Al alloy for example, the near-surface region consists of pure Cu (*i.e.* free of dissolved Al) and Al_2O_3 particles, while below this region only the solid solution exists. The conclusion that we draw from this is that the reaction (oxidation, sulphidation, etc.) is occurring on a well-defined front. We can therefore treat this problem using the same methodology just developed for the evaporation of a reactive species.

Before proceeding, however, it is worthwhile defining the conditions under which internal precipitation is likely to be found. These include:

(a)   a high degree of solubility of the species in the alloy;
(b)   the reaction product is insoluble in the alloy;
(c)   the reacting element in the alloy is present at low concentration (*i.e.* a dilute alloy); and
(d)   the rate of diffusion of the gas species in the alloy is much higher than that of the other reacting species.

The diffusion problem we need to solve is illustrated in Fig. 7.12. We are considering an A–B alloy in which B is the solute. The bulk concentration

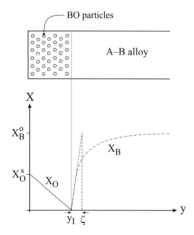

Figure 7.12 A schematic illustration of the concentration profiles for oxygen and solute B in an A–B alloy during internal oxidation. With the internally oxidized region ($y < y_I$) the oxygen concentration is plotted and X_B is zero. Within the unoxidized region ($y > y_I$) the solute concentration is plotted and X_O is zero.

[¶] In the case of vapour bubbles, precipitation is generally found along grain boundaries which, at the temperatures that are generally involved, provide a rapid diffusion path.

of B in the alloy is C_B^o. This is subjected at the surface ($y = 0$) to an oxidizing atmosphere. Due to local equilibrium conditions, the oxygen concentration in the alloy is C_O^s. Based on our previous discussion, we assume that dissolved oxygen diffuses through the layer of material which has already undergone internal oxidation and reacts with B at the interface ($y = y_I$). Thus the boundary conditions in this region (written in terms of fractional concentrations) are:

$$\text{B.C.} \quad \begin{aligned} X_O &= X_O^s, & y &= 0 \\ X_O &= 0, & y &= y_I. \end{aligned}$$

Under pseudo-steady-state conditions, this produces a flux of

$$J_O = \frac{D_O C}{y_I} X_O^s. \tag{7.56}$$

We will assume that the oxide phase formed has the stoichiometry BO.[¶] In this case, an equal but opposite flux of B is required

$$J_B = -D_B C \frac{\partial X_B}{\partial y}\bigg|_{y=y_I} \approx -\frac{D_B C}{\zeta} X_B^o \tag{7.57}$$

where ζ is the thickness (approximate) of the diffusion layer for B in the alloy, ahead of the interface. We would like to calculate the rate of growth of the internally oxidized layer, $\dot{y}_I$. Since we have assumed that all of the B is reacted as the front passes, this just requires a mass balance of the type we have developed several times previously. Suppose the front moves by dy_I in a time interval dt. The amount of oxygen arriving at the front during this time is

$$J_O A\, dt = \frac{D_O C A X_O^s}{y_I}\, dt, \tag{7.58}$$

where A is the surface area of the component being studied. The amount of B converted to oxide in this time interval is equal to

$$C_B^o \cdot A \cdot dy_I = C X_B^o A\, dy_I.$$

Since the number of moles of B and O consumed are equal in this case, we can equate these expressions. Therefore, the rate at which the front moves equals

$$\frac{dy_I}{dt} = \frac{D_O X_O^s}{y_I X_B^o}. \tag{7.59}$$

If the oxide has a different stoichiometry, say B_nO_m, then this result is modified only slightly to

$$\frac{dy_I}{dt} = \frac{m}{n} \frac{D_O X_O^s}{y_I X_B^o}. \tag{7.60}$$

[¶] For other oxides, *e.g.* Al$_2$O$_3$, a simple modification to the relative fluxes is required.

You will recognize this result as giving parabolic oxidation kinetics since, if we integrate over time, the result is

$$y_I = \sqrt{2\,\frac{m}{n}\,\frac{D_O X_O^s}{X_B^o}\,t}. \tag{7.61}$$

Now there is an assumption implicit in this analysis. This is that oxygen diffusion through the layer is the rate-limiting process, and not diffusion of B to the interface. So long as this assumption holds, the thickness of the diffusion layer for B, equal to ζ, will be much smaller than that of the oxidized layer y_I. From eqs. (7.56) and (7.57), since $J_O = -J_B$,

$$\frac{y_I}{\zeta} = \frac{D_O X_O^s}{D_B X_B^o}. \tag{7.62}$$

Therefore, $\zeta \ll y_I$ only if $D_O X_O^s \gg D_B X_B^o$. In other words, internal oxidation requires a dilute alloy (low X_B^o), a soluble gas species (high X_O^s) and a rapidly diffusing gas species. If these conditions are not met, then B will diffuse to the surface where it will meet the oxygen and react to form a continuous surface oxide.

7.6 Further reading

R. J. Borg and G. J. Dienes, *An Introduction to Solid State Diffusion* (Academic Press, San Diego, CA, 1988)

J. R. Poirier and G. H. Geiger, *Transport Phenomena in Materials* *Processing* (TMS-AIME, Warrendale, PA, 1994)

D. R. Gaskell, *An Introduction to Transport Phenomena in Materials Engineering* (Macmillan Publishing, New York, 1992)

7.7 Problems to chapter 7

Note: Relevant phase diagrams are to be found in Appendix C.

7.1 You wish to make a magnetic ferrite $MgFe_2O_4$ by sintering a mixture of 95 wt% Fe_2O_3 powder with 5 wt% MgO, at 1500 °C. Suppose that both materials have an average particle diameter of 5 μm. The density of MgO is 3.5 g/cm^3, while that of Fe_2O_3 is 5.18 g/cm^3.

  (a) Set up the boundary problem you would have to solve to estimate how long it would take to produce a single-phase material. This requires that you:
    – calculate the initial conditions
    – calculate the boundary conditions
    – determine what geometry is best suited to the problem.

Be as precise as possible in describing the conditions. In particular, do not assume that the densities of the two phases are the same. In addition, draw schematic curves, illustrating how the process will proceed, from the initial conditions to its final state at 1500 °C.

(b) Suppose that after sintering for sufficient time to obtain a fully dense homogeneous spinel, this is annealed at 1150 °C. Set up the boundary value for this situation. Make reasonable assumptions about the dimensions involved, and assume that the controlling diffusion coefficient is 10^{-15} m^2/s. Estimate the time required for a new equilibrium to be established.

7.2 The superconductor Nb$_3$Sn can be made by passing Nb wire through a bath of molten Sn at 1000 °C.

(a) Write down the initial conditions and boundary conditions for this problem. Be as precise as possible.

(b) Draw a set of schematic curves which illustrate the formation of Nb$_3$Sn during this process.

(c) Suppose that the growth of the Nb$_3$Sn-phase is controlled by the diffusion of Sn, and that it has a diffusion coefficient of 2×10^{-11} m^2/s at 1000 °C. How long should the Nb wire remain in the Sn bath in order to develop a 100 µm layer of Nb$_3$Sn.

7.3 Suppose you are given a slab of material, which you are told consists of a piece of element A, joined to a piece of element B. You are also told that the A material contains a uniform distribution of A*, a radioactive isotope of A. You are asked to determine as much as you can about diffusion in this couple, and the nature of the phase equilibrium in the system. To do this, you divide the slab into three blocks as shown.

The three blocks are then annealed as follows:

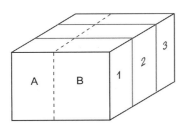

Block 1 is left unannealed.
Block 2 is annealed at 900 °C for 1 hour.
Block 3 is annealed at 1000 °C for 1 hour.

You then measure the radioactivity as a function of distance from the left end of the blocks (*i.e.* the A-rich end). The results are given in Table 7.1.

Table 7.1 Radioactivity as a function of distance from left end of block.

Position from end, y (mm)	Activity (counts/s mg of sample)		
	Block 1	Block 2	Block 3
0	999	1001	1000
100	1002	999	998
150	998	1002	989
175	1001	961	917
180	1000	952	890
185	1003	908	851
190	997	871	819
195	999	828	782
199	1001	781	730
200	2	279	330
205	4	210	281
210	1	151	228
215	2	119	191
220	4	62	147
225	0	27	106
250	1	5	21
300	2	2	2
400	3	2	1

Using this data, determine the following:
(a) Which diffusion coefficients can you estimate?
(b) Determine the value of the diffusion coefficients at each tempera-
    ture. Estimate the activation energies for diffusion in each phase.
(c) What equilibrium concentrations can you determine?
(d) Draw, as precisely as possible, the portion of the phase diagram
    exposed by these three experiments. Be sure to label the phase
    fields and indicate the data points you use. What assumptions
    have you made?
(e) What will be the final position of the interface, assuming that
    the density of the two phases are equal, when diffusion in the
    couple reaches equilibrium? Determine this for both tempera-
    tures.
    (*Hint:* Think carefully about the best way to analyze this data,
        before you begin. Otherwise you may waste a lot of time
        doing unnecessary calculations.)

7.4   (a) Iron powder is sometimes sintered at 1300 °C in the presence of molten copper (a process known as liquid-phase sintering). Set up a model for diffusion in the iron particles by assuming that each particle is coated by a layer of molten copper. Assume spherical particles of 10 μm diameter, and a bulk composition of 15 wt% Cu. As part of your answer:
(i) write down the initial conditions,
(ii) write down the boundary conditions, and
(iii) draw a set of schematic curves for the concentration profiles.
(b) Determine how long it will take for the iron particles to absorb copper up to 80% of the equilibrium amount.

7.5   A commercialized process exists for making metal matrix composites. In this process SiC particles are stirred into an aluminum melt. The molten mixture is then cast. The particles make the composite stiffer and stronger than the un-reinforced alloy. These composites are finding applications in brake rotors and drive shafts for automobiles.

One of the major processing problems is that the aluminum melt can react with the silicon carbide during the stirring process according to:

$$4\,Al + 3\,SiC \rightarrow Al_4C_3 + 3\,Si.$$

This reaction has several detrimental effects which are as follows:[†]

- it produces the reaction product Al_4C_3 at the interface between the reinforcement and the matrix, which could result in a degradation of the reinforcement strength and the interfacial strength;
- the reaction product Al_4C_3 is unstable in some environments, so it increases the corrosion susceptibility of the alloy;
- the reaction increases the silicon content of the alloy and hence the matrix composition of the alloy may be changed significantly if reaction occurs.

From the free energy of the reaction and the activity coefficient of silicon in aluminum, one can determine the silicon content required to stop the reaction (i.e. $\Delta G = 0$ with pure aluminum, silicon carbide and aluminum carbide). At 700 °C, it is a silicon mole fraction of 0.1.

There are three reaction schemes which can occur to be considered in this problem.

(a) At the very start of the reaction, Al_4C_3 crystals form at the interface between SiC and aluminum. Discuss how a crystal grows after the initial nucleation. Show as quantitatively as you can how the silicon concentration in the aluminum varies around these particles. What will limit the rate of growth of the crystals?

[†] D. J. Lloyd, H. Lagace, A. McLeod and P. L. Morris, Microstructural aspects of aluminium–silicon carbide particulate composites produced by a casting method, *Materials Science and Engineering*, **A107**, 73–80 (1989).

(b) After some time, a complete coating of Al_4C_3 will form on each SiC particle. Discuss how this layer will grow, and what will limit the rate of growth. How will the rate of growth be expected to vary with time?

(c) In practice, it is very difficult to prevent the SiC particles from reacting with oxygen in air before they are incorporated into the melt: $SiC + 2O_2 \rightarrow SiO_2 + CO_2$. This reaction coats the particles with a silica film. Consider what would occur when these particles are placed in contact with the aluminum melt. What reaction would occur and what would limit the rate of reaction?

7.6   The process of transient liquid-phase sintering consists in infiltrating a powder product with a liquid, and holding the compact at some high temperature. For example, we might sinter a cobalt powder compact by adding liquid copper at a temperature above the peritectic isotherm (1112 °C), and allowing full solidification to take place isothermally. The amount of liquid is generally less than 10% by volume, and the final composition of the sintered materials will ideally be in the single-phase solid solution region (α-Co in the phase diagram shown in Appendix C).

To get an idea of the kinetics of this process, consider the sintering of a pure Co powder to which pure copper is added to form a liquid film of initial thickness 50 μm at a temperature of 1200 °C. The cobalt grains are much larger than the film thickness. You will need to describe two stages of the sintering treatment.

(a) In the first, rapid stage, the thickness of the liquid film will increase as Co is dissolved in the liquid Cu. You are not asked to estimate the kinetics of this process, but you should determine the thickness of the uniform (in composition and thickness) liquid film at the end of the stage.

(b) In the second, slower stage, isothermal solidification of the uniformly mixed liquid takes place by the diffusion of the Cu into the Co matrix. Develop a model for this process and, from your model, estimate the time required to complete the sintering (*i.e.* to eliminate the liquid phase). The diffusion coefficient of Cu in α-Co is given by

$$2 \times 10^{-4} \exp\left[\frac{-240,000}{RT}\right] \, m^2/s.$$

(c) Using your model, find comparable values of the film thickness at the end of the first stage, and of the time for complete sintering for a temperature of 1367 °C.

    (d) Summarize the assumptions you have made in developing your model.

7.7   Two pure elements, A and B, have negligible terminal solid solubility for each other. However, they do form an intermediate phase A_xB_y with a 10 mol% range of solubility, centred on the equiatomic composition AB. The phase boundaries are nearly vertical between 600 and 900 °C.

    (a) Sketch a set of isothermal free energy vs composition plots which accurately reflects this situation, and a corresponding phase diagram.

    (b) A semi-infinite diffusion couple made up from pure A and B is annealed at two temperatures, 700 and 800 °C, and the rate of growth of the intermediate phase is observed by metallurgical means. In each case, it is parabolic, and the thickness of the layer is given by $\delta = k^*/\sqrt{t}$, where the rate constant k^* is temperature dependent. Develop an expression for k^*.

    (c) If k^* at 800 °C is 5 times that at 700 °C, what is the activation enthalpy for the diffusion process in the phase AB?

7.8   A diffusion couple is prepared by welding a long bar of an A–B alloy $(C_B = 0.033 \, \text{mol cm}^{-3})$ to a long bar of pure metal A $(C_A = 0.100 \, \text{mol cm}^{-3})$. The couple is annealed at 700°C.

    (a) Write the differential equation, initial and boundary conditions that fully describe the changes in composition in this diffusion couple. Use the interdiffusion coefficient, $\tilde{D}$, in your equations. Identify your symbols.

    (b) (i) Write the short-time solution to the boundary-value problem in part (a).

        (ii) Sketch composition profiles in the diffusion couple along the axial centreline in the vicinity of the original interface for several different annealing times.

    (c) Tracer measurements show that $D_A^*/D_B^* = 3$ in the A–B system over the range $0.033 > C_B > 0$. Thermodynamic data show that the A–B system obeys Raoult's law.

        (i) How will the composition profiles differ from those calculated in part (b)?

        (ii) How does the Matano interface move relative to the original interface in fixed space? Use sketches where appropriate.

    (d) Three sets of inert markers are placed in the couple, one at the original interface, one a short distance into the alloy, and one a short distance into pure metal A. Show qualitatively how they move.

(i) if $D_A^* = D_B^*$; (ii) if $D_A^*/D_B^* = 3$.

(e) Will the markers move with iso-concentration fronts?

7.9   A Kirkendall-type experiment was performed in the A–B alloy system. Markers placed at the original interface were found to move with the iso-concentration front $X_A = 0.30$. After a 60-hour anneal, the following data were obtained:

concentration gradient at $X_A = 0.30$, $\partial X_A/\partial y = 300/\text{m}$;
intrinsic diffusivity of A, $D_A = 3.21 \times 10^{-12}\,\text{m}^2/\text{s}$;
intrinsic diffusivity of B, $D_B = 2.01 \times 10^{-12}\,\text{m}^2/\text{s}$.

(a) Calculate the marker displacement with respect to the ends of the piece.

(b) Suppose that the marker displacement of part a) were zero due to porosity.

  (i) Calculate the pore volume per unit cross-sectional area at $t = 30$ hours.

  (ii) In which part of the specimen do you expect the pores to form?

Data: At the temperature of the anneal, the activity coefficient of A in the A–B system is given by $\ln \gamma_A = -0.4X_B^2 + 0.1X_B^3$.

7.10   In an experiment to determine the diffusion coefficient of zinc in argon at 800 °C, a sample of molten zinc is contained in a vertical tube (0.1 m deep) over which an Ar carrier gas (1 atm. pressure) is passed. The concentration of zinc vapour in the exit gas mixture is essentially zero. When steady-state conditions are obtained, the molar flux of zinc vapour through the stagnant gas film of argon in the tube is measured. It is found to be $8.5 \times 10^{-2}\,\text{mole/m}^2\,\text{s}$. Calculate the diffusion coefficient of zinc vapour through argon.

Chapter 8

Mass transport in the presence of convection

Oh, could I flow like thee, and make thy stream
My great example, as it is my theme!
Sir John Denham, *Cooper's Hill*

*In this chapter we continue our study of mass transfer in fluids by relaxing the
assumption of quiescent flow. This raises considerably the level of mathematical
complexity involved in obtaining solutions, to the point where analytical solutions are
not available in most cases. Instead we introduce the concept of a mass transfer
coefficient, analogous to the diffusion coefficient for simple diffusion. We will see that a
wide range of correlations can be made which predict the mass transfer coefficient as a
function of the geometry of the interface, the nature of fluid flow and other material
parameters. These correlations are generally obtained by a combination of
experimentation and numerical simulation. However, in some simple cases rudimentary
analytical models provide useful approximations that we can use. In the latter part of the
chapter we will return to the problem of reactions occurring within fluids. We will see that
analytical solutions are not possible unless the reaction occurs at a well-defined front.*

8.1   Transient diffusion in fluids

In the last chapter we worked mostly with flux equations analogous to Fick's
First Law.[¶] We have noted previously, however, that this form of the diffusion

---

[¶] The one exception was our analysis of interdiffusion, in which we studied transient
diffusion at an interface between two soluble phases.

law is most useful when working on steady-state problems. For transient
problems, it is more convenient to use an equation which contains an explicit
time dependence of concentration. For simple diffusion in dilute solutions, this
is given by Fick's Second Law. Recall that we were able to derive this from
Fick's First Law, by considering an arbitrary concentration profile and applying the principle of mass conservation. We will now repeat this analysis with
two additional features. The first is the inclusion of flow, which we take care of
by replacing the diffusive flux J with the total flux N. The second is the
inclusion of a spatially dependent reaction rate. This latter feature allows us
to treat reactions *within* a phase, as opposed to reactions occurring at a well-
defined interface, as we have considered up to now. In both solids and fluids,
reactions can occur throughout a volume. In the following we will use $\dot{r}_A$ to
denote the amount (in moles) of A which is produced by reaction per unit
volume per unit time. Note that $\dot{r}_A$ is not a vector quantity. It is rather a
volume- and time-dependent scalar, *i.e.* $\dot{r}_A = f(x, y, z, t)$.

We now consider a small volume element in a fluid with dimensions dx, dy,
and dz. In general, a mass balance can be written as:

amount entering − amount leaving + amount produced by reaction

= amount accumulated

Let us first consider flow in one dimension (the y-direction). In this case, the
word equation just written becomes (for the y-direction only)

$$N_i \cdot dx \cdot dz - (N_i + \partial N_i) \cdot dx \cdot dz + \dot{r}_i\, dx \cdot dy \cdot dz = \frac{\partial C_i}{\partial t} \cdot dx \cdot dy \cdot dz.$$

$$(8.1)$$

This reduces readily to

$$\frac{\partial C_i}{\partial t} + \frac{\partial N_i}{\partial y} - \dot{r}_i = 0. \qquad (8.2)$$

Equation (8.2) itself reduces to Fick's Second Law, eq. (3.2), in the absence
of convection ($N_i = J_i$) and internal reaction ($\dot{r}_i = 0$). Moreover, this result
can be easily extended to three dimensions since only the flux is a vector
quantity:

$$\frac{\partial C_i}{\partial t} + \nabla N_i - \dot{r}_i = 0. \qquad (8.3)$$

If we substitute eq. (7.5) and Fick's First Law, eq. (2.2) into this we can
separate the term due to diffusion from that due to drift or convection:

$$\frac{\partial C_i}{\partial t} - \nabla \cdot (D_i \nabla C_i) + v \nabla C_i - \dot{r}_i = 0. \qquad (8.4)$$

where the second term is a dot product (see eq. (3.5)). We can write this equation in two forms depending on whether we wish to work on a molar or a mass basis. In terms of molar units, $C_i = X_i \cdot C$. In condensed phases (*i.e.* solids and liquids), the average concentration C is usually constant. In a gas, C can also be constant if the pressure is uniform. Making this assumption, eq. (8.4) becomes

$$\frac{\partial X_i}{\partial t} - \nabla \cdot (D_i \nabla X_i) + v \nabla X_i - \frac{\dot{r}_i}{C} = 0. \tag{8.5}$$

Similarly, in terms of mass units, in which case it is the overall density ρ that is assumed to be uniform,

$$\frac{\partial X_i^*}{\partial t} - \nabla \cdot (D_i \nabla X_i^*) + v^* \nabla X_i^* - \frac{\dot{r}_i}{\rho} = 0. \tag{8.6}$$

These general equations (eqs. (8.5) and (8.6)) are rather complex in that they treat three-dimensional multicomponent flow, involving internal reactions, drift and convection. Thus, a 'full-blown' solution would often require the use of numerical techniques (*e.g.* finite-element methods). However, in most cases we can simplify things considerably by involving symmetry and reducing problems to one dimension, or because some of the terms can be neglected. Nonetheless, these general equations will form the starting point for many of the problems we will consider as we proceed further.

8.2   Mass transport at a flowing interface

An important class of problems arises when a fluid moves past a stationary body (typically a solid) and mass transfer occurs between the two. We are interested here in determining the transport properties *in the fluid*. There are many examples of this type of problem in which a solid say, is dissolved into a gas or liquid, and the rate of dissolution is controlled by transport in the fluid near the solid surface. Now, before we start, it is important to have a clear picture as to when this is likely to be the case. After all, we know that diffusion is generally much slower in solids than in fluids. Therefore mass transfer will only be controlled by the fluid phase if no diffusion is required in the solid. For example, when a pure solid dissolves into a fluid the only transport required is in the fluid. This is true even when the solid is a compound, provided that the elemental composition remains constant during dissolution. Thus when salt, NaCl, dissolves in water, Na^+ and Cl^- ions dissolve at equal rates. You can contrast this with what happens when a soda lime glass is exposed to water. In this case the Na^+ ions are preferentially leached from the glass and it will be the diffusion of these ions in the solid (*i.e.* the glass) that will control the mass transfer kinetics. Therefore, we must remember that in this section we are treating problems in which the fluid controls transport.

After a relatively short period of time, the overall system establishes a steady-state, in that the rate of dissolution becomes a constant, independent of time. However, at each interface the fluid is constantly being replenished. Thus transient conditions prevail locally.

Let us consider the case of a binary (A/B) fluid flowing past a solid. The transient diffusion process is governed by eq. (8.5) while the flux at the interface is given by eq. (7.14). For the binary case this latter relationship becomes

$$N_A = -CD_A \frac{\partial X_A}{dy} + X_A(N_A + N_B) \tag{8.7}$$

for species A. The solution to these equations is however non-trivial since it requires knowledge of the relationship between the fluid velocity v and the total flux N. So, let us revert momentarily to a problem we do understand, the case for quiescent flow, $v = 0$. In this case the solute diffuses perpendicularly away from the interface. For this we need the special case of eq. (8.5) with $v = 0$ and $\dot{r}_i = 0$. This is of course just Fick's Second Law

$$\frac{\partial X_A}{\partial t} = \nabla \cdot (D_A \nabla X_A).$$

If we assume that D_A is independent of concentration and we consider a planar problem, then this further reduces to

$$\frac{\partial X_A}{\partial t} = D_A \frac{\partial^2 X_A}{\partial y^2}. \tag{8.8}$$

If the interface between solid and fluid establishes a condition of local equilibrium (*i.e.* $X_A = X_A^o$ at $y = 0$), and the concentration far from the interface maintains the bulk concentration, $X_A = X_A^\infty$ as $y \to \infty$, then this equation has the well-known solution

$$\frac{X_A - X_A^\infty}{X_A^o - X_A^\infty} = \text{erfc}\left(\frac{y}{2\sqrt{D_A t}}\right). \tag{8.9}$$

The flux of A leaving the surface is therefore equal to

$$N_A = -CD_A \frac{\partial X_A}{\partial y}\bigg|_{y=0} = C(X_A^o - X_A^\infty)\sqrt{\frac{D_A}{\pi t}}. \tag{8.10}$$

In other words, the surface flux is proportional to the concentration difference times a term which depends on the diffusivity.

Now if a fluid starts to flow along the interface (*i.e.* $v \neq 0$) the rate of mass transfer will increase. This is because fluid flow removes solute from the interface at a faster rate than diffusion. This effectively increases the concentration gradient, and therefore the flux. The solution to this problem is now much more complex. However, the driving force for mass transport remains the same. It is given by the concentration difference, $C_A^o - C_A^\infty = C(X_A^o - X_A^\infty)$. What is needed is a new parameter, similar to the diffusivity, which relates the flux to the overall driving force, in the presence of convection. This new parameter is called a 'mass transfer coefficient', and is defined by an equation of the form

$$N_A = k_D(C_A^o - C_A^\infty). \tag{8.11}$$

It simply states that the flux N_A is proportional to the driving force $(C_A^o - C_A^\infty)$, and we have lumped the rest into the *mass transfer coefficient* k_D. This has units of distance per unit time. However, since transport rates always contain the diffusivity, we would expect k_D to depend either on D/L, where L is some characteristic length, or on $\sqrt{D/t}$, where t is a characteristic time. This expression is directly analogous to Fick's First Law for diffusion, in that it connects a known driving force (the concentration difference across the interface) to a known response (the flux across this interface) through an empirical constant.

8.3   Mass transfer coefficient

The concept of a transfer coefficient is widely used in all fields of transport phenomena, including momentum transfer and heat transfer. In fact there is a clear interrelationship between these since fluid flow influences both heat and mass transfer across interfaces. You may be familiar with parameters such as the Reynolds number Re, which combines the average flow velocity v, the kinematic viscosity[¶] of the fluid ν and some characteristic dimension of the system L,

$$Re = \frac{vL}{\nu} \tag{8.12}$$

Combinations of variables such as this are referred to as *dimensionless parameters*, since they always involve physical quantities grouped together to yield parameters having no physical dimension. By choosing the appropriate dimensionless parameters it is possible to greatly reduce the number of independent variables that have to be dealt with in a complex problem.

[¶] The kinematic viscosity is just the actual viscosity of the fluid normalized by its density, $\nu = \eta/\rho$.

8.3.1   Mass transfer to spheres

To gain some understanding of mass transfer coefficients, and how they are derived, let us look at a specific case by considering a solid spherical particle entrained in a fluid as illustrated in Fig. 8.1. If the fluid flows past the particle by convection then it is constantly being renewed at the surface. Therefore, if the solid is soluble in the fluid each element of the fluid which passes close to the particle will absorb some of the solid material and carry it away.

We start (once again) by considering the case of stagnant (*i.e.* quiescent) flow, $v = 0$. While this problem is somewhat trivial (in fact we have solved others like it already) it does establish the form of the mass transfer coefficient for this case. For a stagnant fluid without internal reaction the governing equation is (from eq. (8.5)).

$$\frac{\partial X_A}{\partial t} = \nabla \cdot (D_A \nabla X_A) = D_A \nabla^2 X_A \tag{8.13}$$

For this case the problem has spherical symmetry for which

$$\frac{\partial X_A}{\partial t} = D \left(\frac{\partial^2 X_A}{\partial r^2} + \frac{2}{r} \frac{\partial X_A}{\partial r} \right), \tag{8.14}$$

where r is the radial coordinate measured from the centre of the sphere. The concentration is fixed by local equilibrium at the solid/fluid interface, while far from the interface the concentration is equal to that in the bulk. Thus,

$$\text{B.C.} \quad \begin{aligned} X_A &= X_A^o, \quad \text{at} \quad r = R \\ X_A &= X_A^\infty, \quad \text{at} \quad r \to \infty, \end{aligned}$$

where R is the sphere radius. This problem should look familiar. It is identical to that of dissolution (or growth) of a solid precipitate which we considered in Chapter 6. The solution is given by eq. (6.10):

$$X_A = X_A^\infty - \frac{R}{r} (X_A^\infty - X_A^o). \tag{8.15}$$

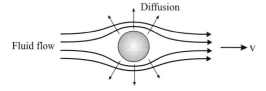

Figure 8.1 A schematic illustration of mass transfer between a solid spherical particle and the surrounding fluid.

Moreover, the flux at the surface of the sphere is

$$N_A = -D_A \left.\frac{\partial C_A}{\partial r}\right|_{r=R} = -D_A C \left.\frac{\partial X_A}{\partial r}\right|_{r=R}$$

$$= \frac{D_A}{R}\left(C_A^o - C_A^\infty\right). \tag{8.16}$$

This result is rather similar to eq. (8.11) which we used to define the mass transfer coefficient. We can see that this new parameter arises naturally from a solution of Fick's Laws in the absence of convection. For this particular case it is given by

$$k_D = \frac{D_A}{R}. \tag{8.17}$$

In keeping with the use of dimensionless numbers to characterize mass transfer, this parameter can be converted into the Sherwood number, which is defined as

$$Sh \equiv k_D \frac{d}{D}, \tag{8.18}$$

where d is the diameter of the sphere $(= 2R)$ and D is the relevant diffusion coefficient (D_A in this case). Thus for the dissolution of a sphere in a stagnant fluid, the Sherwood number Sh is equal to 2.

If the fluid is not stagnant, however, convective flow will increase the rate of mass transfer to the liquid. For simple flow conditions such as laminar flow over simple shapes the mass transfer coefficient can be derived either analytically or using numerical models. For more complex situations the form of the Sherwood number must be determined by experiment. For example, Ranz and Marshall[¶] studied the evaporation of water and benzene drops in flowing air of known velocity (*i.e.* forced convection). They were able to correlate their results with an equation of the form

$$Sh = 2 + 0.6 Re^{1/2} Sc^{1/3}, \tag{8.19}$$

where the Schmidt number Sc is another dimensionless number defined as

$$Sc \equiv \frac{\nu}{D}. \tag{8.20}$$

This relationship has been shown to have general validity over a wide range of conditions (*e.g.* Reynolds number (see eq. (8.12)) varying between 1 and 30,000, and Schmidt number between 0.6 and 3,000). However, it can only be applied when 'forced' convection is operative. In other words, the fluid velocity v past the spheres, which enters the Reynolds number, must be fixed externally.

¶ W. E. Ranz and W. R. Marshall, *Chem. Enging. Progr.*, **48**, 141, 173 (1952)

Convection can also occur 'naturally', driven by gradients in temperature or density throughout the system. For example, when Mn is added to Al alloys the large density difference between these two species causes the Mn particles to sink in the melt as they dissolve, which will naturally affect the rate of dissolution. Similarly, in large refractory furnaces such as glass-melting tanks thermal gradients lead to convective flow. This accelerates the rate of chemical wear of the refractory walls. When natural convection occurs we must use the mass transfer Grashof number, Gr' which correlates the flow driven by density gradients in a fluid of varying composition. This is defined as

$$Gr' = \frac{gL^3\beta'(X_s - X_b)}{\nu^2}.$$

(8.21)

Here g is the gravitational constant (9.81 m/s^2), L the characteristic length (*i.e. d* for a sphere), X_s and X_b are the surface and bulk concentration, respectively (in our problem X_A^o and X_A^∞) and β' is the fractional change of the density of the fluid with concentration (which is what actually drives natural convection)

$$\beta' = \frac{1}{\rho}\left(\frac{\partial\rho}{\partial X}\right)_T.$$

(8.22)

The Grashof parameter can also be used in combination with the Schmidt number to correlate the rate of dissolution of spherical particles when only natural convection is present, for which case

$$Sh = 2 + 0.59(Gr' \cdot Sc)^{1/4}.$$

(8.23)

Similar correlations have been developed which cover both natural *and* forced convection.[¶]

8.3.2   Mass transfer coefficient for other geometries

Clearly mass transfer correlations can be determined, based on both theory and experiment, for a wide range of geometrical configurations. This subject is deeply connected with fluid flow and has been widely covered in texts which address rate phenomena generally. For example, the rate of mass transfer to a fluid passing over a plate will depend on whether the fluid flow is laminar or turbulent. The flow will also vary depending on whether the plate is vertical, horizontal or at some angle. The number of correlations is therefore large. It is essential when using these to understand carefully how they were derived and under what conditions they are valid.

[¶] For a more complete description of this subject consult N. J. Themelis, *Transport and Chemical Rate Phenomena* (Gordon and Breach, Basel, 1995)

Example 8.1: Mass transfer to a well-stirred liquid

Pure hydrogen at 1 atmosphere is in contact with molten iron contained in a circular bath of depth 150 mm. The bath initially contains no dissolved hydrogen. It is stirred vigorously so that the hydrogen concentration is uniform inside the bath. How long will it take for the iron to become half-saturated with hydrogen? Assume that the mass transfer coefficient for hydrogen transfer through the boundary layer at the surface of the bath is 4×10^{-4} m/s. Note that inside the molten iron hydrogen is in atomic form.

From eq. (8.11) we can calculate the flux across the boundary layer at the surface of the bath: $N = k_D(C^o - C^\infty)$. Here C^∞ is the current bulk hydrogen content of the bath and C^o is the concentration at the surface, which we will assume is in equilibrium with the hydrogen atmosphere. If q is the total amount of hydrogen absorbed up to a given time we can write the bulk concentration as $C^\infty = q/(AL)$ where A is the surface area of the bath and L is the depth. The rate at which q increases is just $\dot{q} = N \cdot A$. Therefore the rate of concentration increase is given by

$$\frac{dC}{dt} = \frac{\dot{q}}{AL} = \frac{N}{L} = \frac{k_D}{L}(C^o - C^\infty).$$

Integrating this we find that

$$\ln\left(\frac{C^o - C_i^\infty}{C^o - C^\infty}\right) = \frac{k_D}{L}\,t,$$

where C_i^∞ is the initial bulk concentration (which in this case is zero). Therefore the time required for C^∞ to reach one-half of C^o is

$$t = \frac{0.15}{4 \times 10^{-4}}\ln(2) = 260\,\text{s}.$$

8.4 Models for mass transfer

Although the process of mass transfer to a moving fluid is clearly complex there are some simple models which have been developed. These serve both to give a clearer physical picture of the processes involved and, under certain circumstances, to enable useful calculations of mass transfer to be made. In this section we will consider two such models.

8.4.1 Stagnant film model

When a fluid flows over a surface, a boundary layer always develops near the surface. The flow velocity in this layer is much lower than in the bulk. We can approximate this as a stagnant layer (*i.e.* $v = 0$) out to some depth δ. Within this layer the flux will be given by

$$N_A = \frac{CD_A}{\delta} \frac{(X_A^o - X_A^\infty)}{(\Delta X_B)_{ln}}. \tag{8.24}$$

This follows directly from eq. (7.47) which we developed to treat evaporation in a stagnant well, except that the well depth L has been replaced by the stagnant film thickness δ. From this result we can write the mass transfer coefficient as

$$k_D = \frac{D_A}{\delta(\Delta X_B)_{ln}}. \tag{8.25}$$

This simple expression may appear to offer an attractive approach for estimating mass transfer. However, the stagnant layer thickness is not known a priori. Moreover, it is a function of the same fluid flow parameters as the empirical correlations developed in the previous section. This model therefore offers a conceptual picture of the mass transfer process, but is not of great practical value.

8.4.2 Higbie penetration model

A more useful model is that first developed by Higbie,[¶] also known as the surface renewal model. This model acknowledges that any element of the fluid will only come into contact with a solid surface for a limited period of time, as illustrated in Fig. 8.2.

We now wish to consider such an element. As it approaches the surface the element of fluid will have a uniform composition equal to that of the bulk, C_A^∞. At time $t = 0$, this element comes in contact with the surface, and stays in contact for an exposure time t_e. During this time-period, transient diffusion

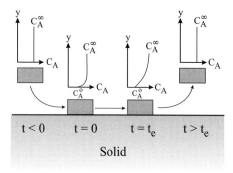

Figure 8.2 A schematic illustration of the Higbie model in which a volume element, initially with the bulk composition, penetrates the stagnant layer, and resides near the surface for a limited period of time t_e.

¶ R. Higbie, *Trans. Amer. J. Chem. Engg.*, **31**, 365 (1935)

occurs across the interface. Since t_e is generally very short, thus allowing limited diffusion into the fluid, we can treat this element as semi-infinite. In other words, we assume that the time t_e is insufficient for the solute to diffuse across the element. This of course requires that the size of the element be a fair bit larger than $\sqrt{Dt_e}$. If we establish local equilibrium at the surface, then the diffusive problem becomes

$$\frac{\partial C_A}{\partial t} = D \frac{\partial^2 C_A}{\partial y^2}, \tag{8.26}$$

subject to

I.C. $C = C_A^\infty$, $t \leq 0$

B.C. $C = C_A^o$, $y = 0$, $0 < t < t_e$

$C = C_A^\infty$, $y \to \infty$, $0 < t < t_e$

This is a problem we have encountered many times by now and its solution (see eq. (8.9)) is just given by

$$\frac{C_A - C_A^o}{C_A^\infty - C_A^o} = \mathrm{erfc}\left(\frac{y}{2\sqrt{D_A t}}\right). \tag{8.27}$$

Moreover, the flux at the surface is given by eq. (8.10). However, what we really need in this case is the flux averaged over the exposure time:

$$\overline{N_A} = \frac{1}{t_e} \int_0^{t_e} N_A \, dt = 2(C_A^o - C_A^\infty) \sqrt{\frac{D_A}{\pi t_e}}. \tag{8.28}$$

By comparing this equation with our definition of the mass transfer coefficient we find that

$$k_D = 2\sqrt{\frac{D_A}{\pi t_e}}. \tag{8.29}$$

Now, as in the previous section, we have derived an expression for k_D which contains an unknown parameter, in this case t_e. However, for simple situations this parameter can be easily estimated.

Consider, for example, the situation of a spherical particle falling in a fluid or a spherical bubble rising in a liquid. In either case the exposure time is simply given by

$$t_e = \frac{d}{v}, \tag{8.30}$$

where d is the bubble or particle diameter and v is the relative velocity of the sphere with respect to its surroundings. Thus,

$$Sh = \frac{k_D d}{D_A} = \frac{2}{\sqrt{\pi}} \sqrt{\frac{vd}{v} \cdot \frac{v}{D_A}} = 1.13 \sqrt{Re \cdot Sc}. \tag{8.31}$$

Although this has a slightly different form than the empirical correlation given earlier (eq. (8.19)), it does yield similar numerical results.

8.4.3 Application to modelling evaporation from molten metal

Let us apply what we have just learned to a common industrial problem. In the production of nodular cast iron it is important to maintain the correct chemistry. Typically, about 1 wt% Si is added to the molten iron to promote graphite formation, while about 0.01 wt% Mg is added to spheroidize the graphite. However, Mg is volatile and evaporates quickly at the surface of the ladle. If the Mg concentration drops too much (below about 0.008 wt%) the cast iron will not develop the necessary microstructure. In practice this problem is overcome by adding additional Mg every so often, a process called 're-inoculation'. The evaporation can only take place at the surface of the melt and is aided by convective flow which continuously brings new iron to the surface (see Fig. 8.3).

Suppose that measurements indicate a flow velocity over the surface of 0.2 m/s, caused by inductive heating of the iron. You wish to know how long it will take before Mg re-inoculation is required. We will assume that once it evaporates the Mg gas is quickly dispersed. Therefore, the partial pressure of magnesium vapour above the vessel is essentially zero. This means that, assuming local equilibrium, the Mg concentration in the alloy in contact with the surface will also be zero, *i.e.* $C_{Mg}^o = 0$. The other necessary data are:

$$D_{Mg\,in\,Fe} = 3 \times 10^{-9}\,m^2/s \text{ (estimated)}$$

$$\rho_{Fe} = 7.0 \times 10^3\,kg/m^3$$

$$H = d = 0.5\,m.$$

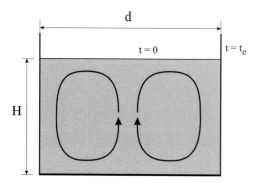

Figure 8.3 An illustration of convective flow in a ladle.

Note that the data for iron apply to the molten state. We treat this using the Higbie penetration model in which cast iron of bulk composition is continuously brought to the surface. The exposure time t_e is approximately $d/2v$, where v is the conductive velocity of the molten iron in the ladle. The flux of Mg to the surface during this exposure is equal to

$$N_{Mg} = k_D(C_{Mg}^o - C_{Mg}^\infty) = -k_D C_{Mg}^b, \tag{8.32}$$

where C_{Mg}^b is the bulk Mg concentration. From the model

$$k_D = 2\sqrt{\frac{D_{Mg}}{\pi t_e}} = \sqrt{\frac{8D_{Mg}v}{\pi d}} \tag{8.33}$$

While these equations tell us the flux, what we really want is the rate of Mg loss from the vessel. This is given by the flux of Mg to the surface, times surface area A and divided by the volume of the vessel V, i.e.

$$\frac{dC_{Mg}^b}{dt} = \frac{N_{Mg}A}{V}. \tag{8.34}$$

Since $V = A \cdot H$, where H is the depth of the vessel,

$$\frac{dC_{Mg}^b}{dt} = \frac{N_{Mg}}{H}. \tag{8.35}$$

We now substitute eq. (8.32) into this to get:

$$\frac{dC_{Mg}^b}{dt} = -\frac{k_D}{H} C_{Mg}^b,$$

or

$$\frac{dX_{Mg}^b}{dt} = -\frac{k_D}{H} X_{Mg}^b.$$

We can integrate from $t = 0$, when $X_{Mg}^b = (X_{Mg}^b)_0$, the original composition (0.01 wt% in this example), to $t = t_r$, the time for re-inoculation, when $X_{Mg}^b = (X_{Mg}^b)_r$ given here as 0.008 wt%. The result is

$$t_r = \frac{H}{k_D} \ln(X_{Mg}^b)\Big|_{(X_{Mg}^b)_0}^{(X_{Mg}^b)_r} \tag{8.36}$$

By substitution we find that the time-interval is equal to 2020 s (about 34 min).

8.5   Quiescent systems containing internal reactions

In the previous chapter, we considered the effect of reactions on diffusion, but only for the case in which the rate of reaction was so rapid that it occurred on a well-defined front? What happens if we try to generalize this for the situation in which the reacting species intermingle as they react. This implies that the rate of chemical reaction is relatively low. We will restrict our attention to quiescent systems in which convective flow is negligible, and to steady-state conditions. Under these conditions, eq. (8.5) reduces to

$$CD_i \nabla^2 X_i + \dot{r}_i = 0, \tag{8.37}$$

where we have implicitly assumed that D_i is a constant. This relationship represents a system of equations, one for each species. So let us further restrict ourselves and consider an A/B binary fluid mixture diffusing in one dimension only, for which we have

$$CD_A \frac{\partial^2 X_A}{\partial y^2} + \dot{r}_A = 0,$$
$$CD_B \frac{\partial^2 X_B}{\partial y^2} + \dot{r}_B = 0. \tag{8.38}$$

This is directly analogous to the problem of Se evaporation considered in Section 7.4.2, except we have relaxed the assumption of a rapid reaction. Now to proceed further in this case, we need to know something about the reaction kinetics. This will depend on the particular reaction with which we are concerned. Suppose that A and B react irreversibly to form a compound according to

$$A + B \rightleftharpoons AB.$$

This reaction is second order, and

$$\dot{r}_A = \dot{r}_B = -k_r C_A C_B = -k_r C^2 X_A X_B, \tag{8.39}$$

where k_r is the reaction-rate constant. Substituting this into eq. (8.38) gives

$$D_A \frac{\partial^2 X_A}{\partial y^2} = k_r C X_A X_B,$$
$$D_B \frac{\partial^2 X_B}{\partial y^2} = k_r C X_A X_B. \tag{8.40}$$

This represents a system of coupled non-linear differential equations which needs to be solved subject to appropriate boundary conditions. Generally, such problems can only be solved using numerical methods. For the case just developed, a relatively straightforward finite-difference routine could be used.

8.6 **Further reading**

N. J. Themelis, *Transport and Chemical Rate Phenomena* (Gordon and Breach, Basel, Switzerland, 1995).

D. R. Poirier and G. H. Geiger, *Transport Phenomena in Materials Processing* (TMS-AIME, Warrendale, PA, 1994).

8.7 **Problems to chapter 8**

Note: Relevant phase diagrams are to be found in Appendix C.

8.1   Moth balls are made from naphthalene, a volatile substance that sublimes over time.

(a) Using the data given below, estimate the mass transfer coefficient of naphthalene from moth balls into stagnant air.

(b) Suppose that the moth balls are left in a drafty room in which air flows past them with an average velocity of 0.1 m/s. Estimate the mass transfer coefficient under these conditions.

(c) Determine the flux of naphthalene vapour away from the surface into stagnant air. If you were unable to calculate a value for the mass transfer coefficient in part (a), use 0.002 m/s.

(d) Develop an expression for the size of the naphthalene balls as a function of time in stagnant air. Calculate the time it will take for the naphthalene to completely sublimate.

Assume that the moth balls have an initial diameter of 10 μm. The diffusion coefficient of naphthalene in air at room temperature is $8 \times 10^{-6} \, \text{m}^2/\text{s}$. Its molar volume is $1.83 \times 10^{-4} \, \text{m}^3$. The equilibrium partial pressure of naphthalene vapour in contact with solid naphthalene is 30 Pa.

8.2   Levitation melting is a technique whereby molten droplets are suspended in a gas stream by an electromagnetic field. This enables containerless processing of small volumes, thus reducing contamination from crucible walls. In a typical case, a piece of iron is melted by heating to 1650 °C to form a droplet 5 mm in diameter. An Ar/H_2 stream is then passed over the melt to reduce the surface oxides. Following this process, the droplet contains 10 ppm of dissolved H. The stream is then changed to pure Ar. Since the electromagnetic field produces a high degree of agitation in the melt it is reasonable to assume that the hydrogen is always uniformly distributed within the droplets.

(a) Determine the mass transfer coefficient for hydrogen gas diffusing into the gas stream.

(b) Based on the thermodynamics data given here, estimate the hydrogen pressure and thus the concentration (moles/m^3) of $H_{2(g)}$ at the surface of the droplet. Do this at the onset of degassing, when the H concentration within the droplet is 10 ppm.

$$H_{2(g)} \rightleftharpoons 2H, \quad K = 2.5 \times 10^{-9} \text{ at } 1650\,°C.$$

(c) Calculate the rate (moles/s) at which H_2 is removed from the droplet at the onset of outgassing. (If you did not get reasonable answers to parts (a) or (b) use $k_D = 1\,\text{m/s}$ and $C^\circ = 10^{-6}$ moles/m^3.)

(d) Convert your answer for (c) into the rate at which the dissolved H concentration changes per unit time (ppm/s).

Data: Velocity of Ar stream, $v = 3\,\text{m/s}$
$D_{H_2 \text{ in Ar}} = 10^{-3}\,\text{m}^2/\text{s at } 1650\,°C$
Density of iron, $\rho = 7.8 \times 10^3\,\text{kg/m}^3$
Molar weight of iron, $W_m = 0.56\,\text{kg/mole}$.

8.3   You are concerned with the evaporative loss of magnesium from cast-iron melts intended for spheroidal graphite iron production. In a laboratory study of the evaporation of a dilute volatile solute (magnesium) from a crucible containing a liquid solvent metal (iron–carbon eutectic in this case), two experimental conditions are employed. In the first (A), a strong DC magnetic field is imposed on the crucible. This is an electromagnetic 'clamp' with the effect of suppressing convection. In a second (B), free convection is permitted.
The following dimensions and quantities are known or measured:
• depth of liquid in crucible: 10 mm
• inside diameter of crucible: 20 cm.
• diffusion coefficient of magnesium in liquid iron: $3 \times 10^{-9}\,\text{m}^2/\text{s}$
• v^*, the observed velocity of liquid metal on the surface of the freely convecting melt (case B): 0.05 m/s.

(a) Determine and compare the times required for the reduction of the *average* Mg composition of the melt from 0.1 mol.% to 0.01 mol.%, for the 'clamped' (A) condition and for the free-convection (B) condition.

(b) Describe the assumptions implicit in the model you have chosen to represent the convective mass transfer for each case.

(c) Laboratory experiments can provide certain information about corresponding industrial processes. Discuss the methods used to compare the two.

8.4   Magnesium is used to desulphurize molten iron from the blast fur-
        nace, and also to produce nodular iron. Magnesium boils at 1107 °C,
        and therefore vaporizes when injected subsurface into iron, which is
        typically at 1250 °C. To be used effectively in the iron, the magnesium
        vapour bubbles rising in the iron must dissolve before they reach the
        bath surface. This depends on two factors: the equilibrium solubility
        of magnesium in iron; and the kinetics of magnesium vapour dissolu-
        tion into iron from the bubble.

        Calculate how long it takes for a 10 mm diameter spherical bubble of
        pure magnesium vapour to dissolve when rising through molten iron
        with a velocity given by Stokes Law:

$$U_{\mathrm{b}} = \frac{d_{\mathrm{b}}^2 g (\rho_{\mathrm{Fe}} - \rho_{\mathrm{v}})}{18\eta}$$

        where U_{b} bubble rising velocity, m/s
                d_{b} bubble diameter, m
                g gravitational acceleration, 9.81 m/s^2
                ρ_{Fe} iron density, 7000 kg/m^3
                ρ_{v} vapour density
                η viscosity of iron, 9×10^{-3} kg/m s

        As the bubble rises the ferrostatic head will decrease, making the
        bubble expand, thus counteracting the decrease in size from dis-
        solution; ignore any change in ferrostatic head and assume that the
        pressure is constant at 170 kPa (1 metre depth).

        The solubility of magnesium in iron is given by Henry's Law: at
        equilibrium, the amount that can dissolve is proportional to the
        magnesium vapour pressure. At the temperature of the iron, 1250 °C,
        the constant of proportionality, Henry's Constant, is 0.7 wt% Mg in
        solution per atmosphere of Mg vapour.

        Assume at the time under consideration that there is already
        0.03 wt% Mg dissolved (86.4 mole/m^3) and that this is not changed
        during the rise of this bubble. Take the diffusivity of magnesium in
        molten iron to be 3×10^{-9} m^2/s.

8.5   Chemical vapour deposition (CVD) processes are used to make cera-
        mic coatings on metallic substrates for improved corrosion resistance,
        abrasion resistance or refractory properties. For example, TiC can be
        produced by introducing a mixture of $TiCl_4$ vapour and CH_4 to the
        substrate by the reaction:

$$TiCl_4 + CH_4 \rightleftharpoons TiC + 4HCl$$

$$\Delta G^{\circ}_{1400\,\mathrm{K}} = -60\,\mathrm{kJ/mol}.$$

In an experimental CVD apparatus, a steel plate $0.1\,\text{m} \times 0.1\,\text{m}$ is exposed to a stream of argon containing 0.01 mole fraction each of $TiCl_4$ and CH_4 at 1400 K. The gas velocity is 1 m/s, and assume this is fast enough that the concentration in the stream of the HCl produced is zero. Assume that the diffusivities of all gas species are $5 \times 10^{-4}\,\text{m}^2/\text{s}$.

(a) Sketch the concentration profiles for $TiCl_4$, CH_4 and HCl in the vicinity of the plate.

(b) Is the mass transfer rate higher at the leading edge or trailing edge of the plate, and why?

(c) Calculate the interfacial concentrations of $TiCl_4$, CH_4 and HCl in this particular case.

(d) Calculate the mass transfer coefficient at the centre of the plate.

(e) Calculate the rate of mass transfer at the centre of the plate.

8.6   Nitrogen has a small but significant solubility in liquid steel. At equilibrium at 1600 °C, 443 ppm N dissolves in pure iron when exposed to pure nitrogen at one atmosphere pressure. It dissolves according to Sievert's Law, *i.e.*

$$\tfrac{1}{2}N_2 \rightleftharpoons \underline{N}.$$

The equilibrium constant for this reaction at 1600 °C is:

$$K = \frac{C_{\underline{N}}}{p_{N_2}^{0.5}} = 0.716\,\frac{\text{mol}}{\text{m}^3(\text{Pa})^{0.5}}.$$

In continuous casting of steel, the steel ladle is positioned over the tundish which is used to maintain a constant ferrostatic head or pouring speed and to split the metal into several streams or 'strands'. In a particular 'mini-mill', 150 tonne heats of steel are produced in electric arc furnaces and are cast into square billets in a three-strand continuous casting operation. (Note: 1 tonne = 1 metric ton = 1000 kg.) Assume the 150 tonne heat at 1600 °C is cast in 1/2 hour into three streams of 30 mm diameter each and that the fall height is 1 m. If the casting streams are unshrouded, they are exposed to nitrogen in the air. You wish to calculate the amount of nitrogen pick-up during the 1 m fall from tundish to mould without any protective shroud.

Assume: $D_{N_2-O_2} = 10^{-2}\,\text{m}^2/\text{s}$ at 1600 °C;

$$D_{\underline{N}\,\text{in Fe}} = 4.0 \times 10^{-9}\,\text{m}^2/\text{s} \text{ at } 1600\,^{\circ}\text{C}.$$

(a) Calculate the mass transfer coefficient for nitrogen dissolution assuming that the velocity in the stream is uniform in all parts of the stream.

(b) Calculate the rate of nitrogen pick-up [mol/s] and change in concentration [mol/m^3 of steel]. If you could not determine a mass transfer coefficient in part (a), use 5×10^{-4} m/s.

Hint: The fact that the solubility is very low simplifies the problem.

8.7   Thermodynamically, calcium is a very strong deoxidizer of molten steel. The equilibrium constant for the product of dissolved oxygen and calcium contents at 1600 °C is:

$$CaO \rightleftharpoons \underline{Ca} + \underline{O}$$

$$K = (\%Ca)(\%O) = 10^{-10} \text{ (units of wt\%)}.$$

However, it is difficult to dissolve calcium in the steel because calcium is very volatile (its boiling point is 1492 °C), compared with a typical steelmaking temperature of 1600 °C. The usual mechanism of deoxidation is precipitation of CaO on small (<100 μm) solid-oxide-including particles which are generally present in any steel melt.

(a) Assuming that an excess amount of oxygen and calcium are present in the melt, and assuming that equilibrium is established at the inclusion interface, sketch the oxygen and calcium concentration profiles in the vicinity of the inclusion.

(b) Assuming that oxygen and calcium have equal diffusivities in steel (10^{-8} m^2/s), develop a general expression for the rates of deoxidation [d(%$\underline{O}$)/dt] in terms of the following variables only:
   - the liquid phase mass transfer coefficient, k_D,
   - the bulk concentrations of oxygen and calcium,
   - the equilibrium constant, K, and
   - the number of inclusions per unit volume, n_i.

(c) Calculate the initial rate of deoxidation for a steel containing 0.002 wt% Ca, and 0.02 wt% O and 10^{12} spherical inclusions per cubic metre of 50 μm diameter in a 150 tonne teeming ladle. What approximations to the general case (part (b)) can be made to obtain a simpler expression?

Chapter 9

Advanced topics

For who would lose,
Though full of pain, this intellectual being,
Those thoughts that wander through eternity,
To perish, swallowed up and lost
In the wide womb of uncreated night,
devoid of sense and motion?
John Milton, Paradise Lost

In this chapter we introduce several advanced topics. The first deals with different approaches that can be taken to develop overall mass balances within a vessel. This can be used to simplify the analysis of a complex process, as we will see. We then turn to the problem of multi-phase resistances. Up to now we have been able to make assumptions about which of two phases in contact dominates the mass transfer kinetics across the interface between them. However, there are situations in which both phases may contribute to the overall mass transfer, at least over a limited range of conditions. We will develop an approach for dealing with this situation. Finally, we will consider a specific case involving topochemical reactions in porous solids.

9.1 Overall mass balance

We have used mass balances extensively throughout this book. Most often we have used them to develop an expression for the motion of an interface separating two reacting phases. In fluids, however, this process can become quite complex. Consider, for example, what happens when a gas stream is passed through or over a bed containing some reactive species. The reaction occurs slowly as the gas stream passes. We can simplify the solution if we understand something about the nature of flow in the reacting vessel and the relative rates of flow vs reaction. The most general way to treat such a problem involves developing a mass balance over an infinitesimal volume element at some arbitrary position in the reactor. We can then integrate this solution over the entire volume (assuming that we can determine the appropriate boundary conditions at each location in the vessel). This is clearly complex, only amenable to solution using numerical methods.

In many systems, we make assumptions that remove much of the complexity. For example, if the rate of mixing of the reactive species is very rapid compared with the rate of reaction, then the concentration will be uniform throughout the vessel. This would be the case in a well-stirred tank reactor with low throughput. A more realistic scenario involves a situation in which the reactants flow through a vessel in one direction only. Consider, for example, your garden hose when you turn it on during a hot day. The first water to come out is the warm water that has been sitting in the hose, and only when the water that was in the supply line reaches the end of the hose does the water cool. This is an example of plug-flow. In this case, the problem becomes one-dimensional. Many industrial reactors lie somewhere between these two extremes of complete mixing and plug-flow and need to be treated numerically. The methods required are well developed but are beyond the scope of this text.

Let us consider an example of the plug-flow type involving a flue gas scrubber. Many chemical processes produce SO_2 gas as an unwanted by-product. This must be removed from the effluent gas stream before it is vented to the atmosphere. This can be accomplished by passing the gas through a reactor containing an aqueous NaOH solution. The reaction involved is

$$2NaOH + 2SO_2 + \tfrac{1}{2}O_2 \rightarrow 2NaSO_3 + H_2O.$$

The scrubber is essentially a cylindrical vessel. The effluent gas rises through the vessel and so the concentration of SO_2 is uniform across the diameter of the vessel and varies only with height z (see Fig. 9.1). To enhance the contact area between the gas and the liquid, the reactor is usually filled with ceramic or plastic shapes (such as rings or saddles), and is referred to as a packed bed. We can solve the problem using a mass balance between the decrease in SO_2 concentration on going from z to $(z + dz)$ and the rate of reaction between the gas and the liquid. In order to simplify the analysis, but also consistent with experimental conditions, we will make several further assumptions, namely that:

(i)   the effluent gas is dilute in SO_2,
(ii)   there is an excess of O_2 and NaOH (in which case the concentration does not vary with height in the vessel), and

(iii)   mass transport at the gas–bubble/liquid interface is the rate controlling process.

The mass balance which we will set up now is very similar in form to that used in the derivation of Fick's Second Law.

The SO_2 flux in the gas stream is $C \cdot v$, where C is the overall SO_2 concentration at any given height in the scrubber and v is the molar average gas velocity. Across an element of height dz, the amount of SO_2 removed per unit time is

$$[C(z + dz) - C(z)]vA = vA \frac{\partial C}{\partial z} \cdot dz,$$

where A is the cross-sectional area of the vessel. This must be equal to the reaction rate $\dot{q}$, i.e.

$$\dot{q} = vA \frac{\partial C}{\partial z} \cdot dz. \tag{9.1}$$

Now this relationship is valid at a macroscopic level. However, the reaction only takes place at the gas-bubble/liquid interface where SO_2 reacts with NaOH to form $NaSO_3$. The equilibrium constant for this reaction is

$$K = \frac{C_{NaSO_3}^2 \cdot C_{H_2O}}{C_{NaOH}^2 \cdot C_{SO_2}^2 \cdot C_{O_2}^{1/2}}. \tag{9.2}$$

Assuming local equilibrium at the gas-bubble/liquid interface the $NaSO_3$ concentration there will be given by

$$C_{NaSO_3} = K^{1/2} \frac{C_{NaOH} \cdot C_{O_2}^{1/4} \cdot C_{SO_2}}{C_{H_2O}^{1/2}}. \tag{9.3}$$

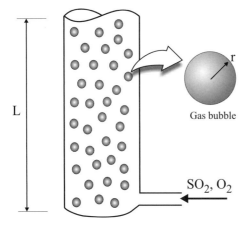

Figure 9.1 A schematic illustration of a flue gas scrubber.

However, because of the assumptions we made earlier, C_{NaOH}, C_{O_2} and C_{H_2O} are all constant as the bubbles rise through the reactor. We can therefore write

$$C_{NaSO_3} = k_p C_{SO_2}, \tag{9.4}$$

where

$$k_p = K^{1/2} \frac{C_{NaOH} \cdot C_{O_2}^{1/4}}{C_{H_2O}^{1/2}} \tag{9.5}$$

is a constant. Now, in order to avoid confusion between the two species SO_2 and $NaSO_3$ and to avoid a proliferation of unwieldy subscripts in what follows, it will be useful to adopt different symbols for the concentration of each species. We will therefore use C to refer to the SO_2 concentration in the gas phase and ζ to refer to the $NaSO_3$ concentration in the liquid phase. This is consistent with the use of C in eq. (9.1). We can now rewrite eq. (9.1) in terms of the $NaSO_3$ concentration on the liquid side of the interface as:

$$\dot{q} = \frac{vA}{k_p} \frac{\partial \zeta^*}{\partial z} \cdot dz, \tag{9.6}$$

where the superscript * indicates that the $NaSO_3$ concentration is that which is in equilibrium with the SO_2 gas concentration.

We let a equal the interfacial area per unit volume in the reactor. (This is a parameter of the system which needs to be measured experimentally). At this level, the reaction rate in the volume element is equal to the flux of $NaSO_3$ through the surface boundary layer times the total interfacial area, i.e.

$$\dot{q} = k_D(\zeta^\infty - \zeta^*)a \cdot A \, dz, \tag{9.7}$$

where ζ^∞ is the $NaSO_3$ concentration in the bulk of the liquid. Because of the complex geometry involved, the mass transfer coefficient k_D would also need to be determined through experiment. Equating (9.6) and (9.7) gives

$$\frac{\partial \zeta^*}{\partial z} \cdot \frac{v}{k_p} = k_D(\zeta^\infty - \zeta^*)a. \tag{9.8}$$

We now collect the concentration terms on one side of the equation and integrate from the inlet to the vessel (set to $z = 0$). The result is

$$\int_{\zeta_i^*}^{\zeta^*} \frac{d\zeta^*}{\zeta^\infty - \zeta^*} = k_D k_p \frac{a}{v} \int_0^z dz$$

or

$$\ln\left(\frac{\zeta^\infty - \zeta^*}{\zeta^\infty - \zeta_i^*}\right) = \frac{k_D k_p}{v} az. \tag{9.9}$$

Here, ζ_i^* is the initial $NaSO_3$ concentration on the liquid side of a bubble as it enters the scrubber. This equation provides the basis for the required experimental measurement of k_D and a. Usually a small-scale column is set up with the same packing, SO_2, O_2, and NaOH concentrations and gas velocity as the full-scale unit. For given initial and final SO_2 concentrations and knowing v and z, the product $k_D \cdot a$ can be deduced. Using the plug-flow assumption this parameter can now be applied to the large-scale column.

We would like to know the rate at which SO_2 is removed by a scrubber of height L. We get this, the total scrubbing rate $\dot{q}_t$, directly from eq. (9.1) integrated over the height,

$$\dot{q}_t = vA \int_0^L \frac{\partial C}{\partial z}\, \mathrm{d}z = vA(C_L - C_i) = \frac{vA}{k_p}\left(\zeta_L^* - \zeta_i^*\right) \tag{9.10}$$

where C_L and ζ_L^* are the SO_2 concentration in the effluent and the $NaSO_3$ interfacial concentration respectively at the top of the scrubber. We see that the total scrubbing rate depends on both the gas velocity and C_L, both of which can be altered. In most cases, the value of C_L will be fixed either by environmental regulation or company policy. So, let us eliminate velocity from this result. We do this by solving eq. (9.9) for the velocity, with $z = L$. Thus,

$$v = \frac{k_D k_p L}{\ln\left(\dfrac{\zeta^\infty - \zeta_L^*}{\zeta^\infty - \zeta_i^*}\right)} \qquad a = \frac{k_D k_p L}{\ln\left(\dfrac{\zeta^\infty - k_p C_L}{\zeta^\infty - k_p C_i}\right)}\, a. \tag{9.11}$$

This tells us the maximum stream velocity we can use in order to reduce the SO_2 concentration from C_i to C_L. Substituting this back into eq. (9.10) gives

$$\dot{q}_t = k_D k_p a L A \frac{C_L - C_i}{\ln\left(\dfrac{\zeta^\infty - k_p C_L}{\zeta^\infty - k_p C_i}\right)}. \tag{9.12}$$

Note that $a \cdot LA$ is just the total interfacial area in the vessel. This analysis can be used in the fundamental design of a gas treatment plant. For a given value of $k_D \cdot a$ and a maximum permissible value of C_L, the column height L is fixed. The total scrubbing rate and the column area (or total area of parallel columns) can be determined. In practice the situation is more complex because the aqueous solution is usually recirculated (from top to bottom) and its NaOH concentration changes. A complete analysis would need to incorporate this effect.

9.2   Multi-phase resistances

Up to now, we have generally tackled problems in which transport in one of the two adjoining phases clearly dominates the process. However, it is not always possible to determine, before starting a problem, which phase will control the overall kinetics. Moreover, in some cases, transport in both phases may be important. For example, in the gas scrubber which we discussed in the previous section either the SO_2, O_2 or NaOH concentrations could control the reaction. We treated SO_2 control by assuming a dilute concentration of this species. In many slag–metal reactions for the removal of elements such as S and P from steel the controlling kinetics can be in either the steel or the slag. We will therefore now develop a framework which will enable us to calculate the overall transport kinetics in a multi-phase system.

Let us consider two phases α and β which meet at a plane interface (see Fig. 9.2). These phases could be solid, liquid or gas. For simplicity, we will consider the diffusion of dilute solute A from the α-phase into the β-phase.

$$A_{(\alpha)} \rightleftharpoons A_{(\beta)}.$$

We will assume that the chemical reaction at the interface is exceedingly rapid, so that equilibrium is established locally at the interface. The equilibrium constant for the reaction is

$$K = \frac{(C_A)^*_\beta}{(C_A)^*_\alpha} \tag{9.13}$$

where the superscript * is used to denote the equilibrium concentration. We start with a uniform composition in each phase, $(C_A)^o_\alpha$ and $(C_A)^o_\beta$, which will generally not be equal to the equilibrium value. Local equilibrium is established instantaneously at the interface (see Fig. 9.2).

The flux is continuous across the interface. Therefore

$$
\begin{aligned}
N_A &= k_{A\alpha}[(C_A)^o_\alpha - (C_A)^*_\alpha] \\
&= k_{A\beta}[(C_A)^*_\beta - (C_A)^o_\beta],
\end{aligned}
\tag{9.14}
$$

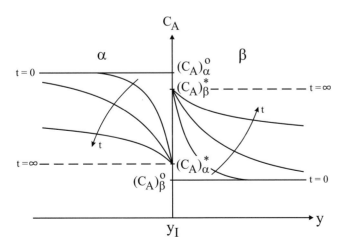

Figure 9.2 A schematic illustration of partitioning between two phases both of which have some solubility for the solute A.

where $k_{A\alpha}$ and $k_{A\beta}$ are the mass transfer coefficients for A in the α- and β-phases, respectively.

Equations (9.13) and (9.14) actually constitute a set of three equations with three unknown parameters: N_A, $(C_A)^*_\alpha$ and $(C_A)^*_\beta$. To solve these, we substitute eq. (9.13) into the first part of eq. (9.14) to obtain

$$N_A = k_{A\alpha}\left[(C_A)^o_\alpha - \frac{(C_A)^*_\beta}{K}\right]. \tag{9.15}$$

In addition, the second part of eq. (9.14) can be rearranged to give

$$(C_A)^*_\beta = (C_A)^o_\beta + \frac{N_A}{k_{A\beta}}. \tag{9.16}$$

Substituting this into eq. (9.14) gives

$$N_A = k_{A\alpha}\left[(C_A)^o_\alpha - \frac{(C_A)^o_\beta}{K} - \frac{N_A}{Kk_{A\beta}}\right]. \tag{9.17}$$

This eliminates all unknown variables except N_A. Solving for this gives

$$N_A = k_{ov}\left[(C_A)^o_\alpha - \frac{(C_A)^o_\beta}{K}\right], \tag{9.18}$$

where the overall mass transfer coefficient is given by

$$\frac{1}{k_{ov}} = \frac{1}{k_{A\alpha}} + \frac{1}{Kk_{A\beta}}. \tag{9.19}$$

It is clear from eq. (9.19) that the phase with the lowest mass transfer coefficient will tend to dominate. However, this is tempered by the equilibrium constant. Thus a high value of K, which drives the solute towards the β-phase, increases the effective mass transfer coefficient in that phase, and vice-versa.

Two limiting cases can be discussed from this result. When transport in the α-phase is rapid, such that $k_{A\alpha} \gg K \cdot k_{A\beta}$, then

$$N_A = k_{A\beta}[K(C_A)^o_\alpha - (C_A)^*_\beta] \tag{9.20}$$

and mass transfer is controlled in the β-phase. In this case, transport in the α-phase is rapid so that $(C_A)^o_\alpha \approx (C_A)^*_\alpha$. Since $(C_A)^*_\alpha = (C_A)^*_\beta/K$, eq. (9.20) simplifies to

$$N_A = k_{A\beta}[(C_A)^*_\beta - (C_A)^o_\beta]. \tag{9.21}$$

This is, of course, just the result we would have obtained had we assumed full control by the β-phase from the beginning.

The opposite result is obtained if we assume that $k_{A\alpha} \ll Kk_{A\beta}$, in which case the α-phase dominates the kinetics. This situation arises often because in metal-refining operations (either gas–liquid metal or slag–liquid metal) one deliberately picks a system with a large K. For example, if β is the slag and

α is the metal, the affinity of the slag for S, P and other impurities is such that K is at least 1,000. This pushes control to the metal side, regardless of the values of $k_{A\alpha}$ and $k_{A\beta}$, which simplifies the analysis for many systems.

However, situations may arise in which both phases contribute significantly to mass transfer. Then the full solution given by eqs. (9.18) and (9.19) should be used.

9.3   Topochemical reaction kinetics

We will conclude this chapter by considering a specific kind of multi-phase resistance problem involving reactions between solids and gases. There are numerous examples of material processes in which the raw materials consist of porous solids, typically in the form of pellets. These are then converted to a second solid phase by way of a gas reaction. The reaction proceeds with a well-defined front from the surface of the pellet toward the centre. We call this a topochemical reaction.

One example of this kind of process involves the reduction of iron ore, which occurs in stages from Fe_2O_3 to Fe_3O_4 to FeO and finally to Fe. The first stage involves a reaction of the form:

$$3Fe_2O_3 + CO \rightarrow 2Fe_3O_4 + CO_2.$$

Another example involves the roasting of sulphide ores, such as copper:

$$CuS + \tfrac{3}{2}O_2 \rightarrow CuO + SO_2.$$

9.3.1   Reduction of oxide pellets

We will consider the reduction of Fe_2O_3 pellets, illustrated schematically in Fig. 9.3, in which CO and CO_2 gases must diffuse through the porous Fe_3O_4 layer. Once the reaction has proceeded some distance into the pellet we can identify five sequential processes that are required for this reaction to proceed.

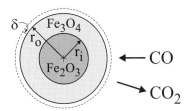

Figure 9.3 A schematic illustration of the reduction of an iron-ore pellet. The reaction occurs at a well-defined front, and both the reactant and product gases must diffuse through the porous reacted layer.

These are:

(1) Diffusion of the reactant gas through the boundary layer near the pellet surface ($r_o < r < r_o + \delta$).
(2) Diffusion of the reactant gas through the reacted layer ($r_i < r < r_o$).
(3) Reaction at the interface between reactants and products (at $r = r_i$).
(4) Diffusion of the product gas through the reacted layer.
(5) Diffusion of the product gas through the boundary layer.

The concentration profile we expect is shown schematically in Fig. 9.4. Here δ is the width of gaseous boundary layer at the pellet surface ($\delta \ll r_o$). A and B represent the reactant and product gas, respectively. Thus for Fe_2O_3 reduction, A is CO and B is CO_2.

We can now write a series of rate equations for each species. For the reactant gas

$$\dot{q}_A = k_D \cdot 4\pi r_o^2 (C_{A,b} - C_{A,s}) \tag{9.22}$$

through the boundary layer, with a mass transfer coefficient of k_D. Here, $C_{A,b}$ and $C_{A,s}$ are the concentrations of the reactant at the edge of the boundary layer and the particle surface, respectively (see Fig. 9.4). Similarly,

$$\dot{q}_A = 4\pi r^2 D_{A,eff} \frac{\partial C_A}{\partial r} \tag{9.23}$$

through the reacted layer. The diffusion coefficient $D_{A,eff}$ represents gaseous transport of A through a porous solid. We will return to look at this in more detail shortly. For now we just treat this as we would any diffusion coefficient. We know that since all of the reaction is occurring at the topochemical interface ($r = r_i$) the flow rate of the gas $\dot{q}_A$ through the region $r_i < r < r_o$ must be constant. (This is essentially the assumption of a pseudo-steady-state that we have used frequently in earlier chapters). We can therefore

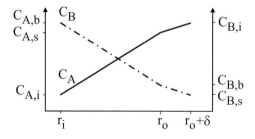

Figure 9.4 The concentration profiles for reduction of an oxide pellet are shown schematically. There are three regions – a boundary layer in the gas phase adjacent to the particle surface, the reacted layer through which both reactant and product gases must diffuse, and the unreacted core.

integrate eq. (9.23):

$$\int_{r_i}^{r_o} \frac{dr}{r^2} = \frac{4\pi D_{A,eff}}{\dot{q}_A} \int_{C_{A,i}}^{C_{A,s}} dC_A, \tag{9.24}$$

where $C_{A,i}$ is the reactant concentration at the topochemical interface. From this we find that

$$\dot{q}_A = 4\pi \frac{r_o r_i}{r_o - r_i} D_{A,eff}(C_{A,s} - C_{A,i}). \tag{9.25}$$

We now need to consider the kinetics of the reaction itself at $r = r_i$. Since this is a sharp interface the solid species are pure Fe_2O_3 and pure Fe_3O_4. Thus the reaction rates, expressed as fluxes, are

$$N_A = k_r C_{A,i}$$

for the forward reaction, and

$$N_A = -k_r' C_{B,i} = -k_r'(C - C_{A,i})$$

for the reverse reaction, where C is the overall gas concentration. The net reaction flux is therefore

$$N_A = k_r C_{A,i} - k_r' C_{B,i} = k_r \left(C_{A,i} + \frac{C_{A,i} - C}{K} \right), \tag{9.26}$$

where

$$K = \frac{k_r}{k_r'} = \frac{C_{B,i}}{C_{A,i}} \tag{9.27}$$

is the equilibrium constant for the reaction. If we multiply the flux by the area of the interface we get a flow rate compatible in form to those in eqs. (9.22) and (9.25). Thus

$$\dot{q}_A = 4\pi r_i^2 k_r \left(C_{A,i} + \frac{C_{A,i} - C}{K} \right). \tag{9.28}$$

We now have three flow equations (9.22), (9.25) and (9.28). Conservation of mass suggests that $\dot{q}_A$ is the same for each of these. We therefore have only three unknown parameters $(\dot{q}_A, C_{A,s}, C_{A,i})$ and so these equations can be solved. To do this, we invert eq. (9.22) to give

$$\frac{C_{A,b} - C_{A,s}}{\dot{q}_A} = \frac{1}{k_D \cdot 4\pi r_o^2}, \tag{9.29}$$

and eq. (9.25) to give

$$\frac{C_{A,s} - C_{A,i}}{\dot{q}_A} = \frac{r_o - r_i}{r_o \cdot r_i} \cdot \frac{1}{4\pi D_{A,eff}}, \tag{9.30}$$

We now relate $C_{A,i}$ to the equilibrium concentration $C_{A,e} = C_{B,e}/K$. Since $C = C_{A,e} + C_{B,e} = C_{A,e}(1 + K)$, the term in parentheses in eq. (9.28) becomes

$$C_{A,i} + \frac{C_{A,i}}{K} - \frac{C}{K} = C_{A,i}\left(1 + \frac{1}{K}\right) - C_{A,e}\left(1 + \frac{1}{K}\right). \tag{9.31}$$

Thus, eq. (9.28) can be converted to

$$\frac{C_{A,i} - C_{A,e}}{\dot{q}_A} = \frac{1}{4\pi r_i^2 k_r\left(1 + \dfrac{1}{K}\right)}. \tag{9.32}$$

If we now add eqs. (9.29), (9.30) and (9.32) all of the concentration terms cancel except $C_{A,b}$ and $C_{A,e}$, both of which are known. Therefore:

$$\frac{C_{A,b} - C_{A,e}}{\dot{q}_A} = \frac{1}{k_{ov}}\frac{1}{4\pi r_o^2}, \tag{9.33}$$

where

$$\frac{1}{k_{ov}} = \frac{1}{k_D} + \frac{r_o - r_i}{\dfrac{r_i}{r_o} D_{A,eff}} + \frac{1}{\left(\dfrac{r_i}{r_o}\right)^2 k_r\left(1 + \dfrac{1}{K}\right)} \tag{9.34}$$

is the overall mass transfer coefficient. Of course, we can easily invert eq. (9.33) to give the flow rate at the surface of the pellet:

$$\dot{q}_A = k_{ov}4\pi r_o^2(C_{A,b} - C_{A,e}). \tag{9.35}$$

Now all of this applies to the reactant gas A. We could also do the same analysis for the product gas B. Since these two processes are dependent on one another the overall kinetics of the process will be determined by the species giving the smallest flux.

9.3.2   Gas diffusivity in porous solids

In the previous section we assigned an 'effective' diffusion coefficient to the gas moving through the porous reacted layer. Why is this different from gas diffusion in an open space? Well, there are two possible reasons. The first, and simplest, is that the effective path length is longer for a porous body. We use the diffusivity to relate the concentration gradient to the flux. In a porous solid, the gas cannot travel directly down the concentration gradient. It must take a more tortuous path around the particles. In addition, the cross-sectional area available is also reduced by the volume fraction of porosity in the solid ω. The effective diffusivity can therefore be written as

$$D_{eff} = D\frac{\omega}{\tau}$$

where D is the diffusivity of the same species in open space and τ is called the 'tortuosity'. For porous solids, the porosity ω generally ranges from about 0.2 to close to 1. The tortuosity is always larger than unity. It will be close to 1 for a highly porous solid (*e.g.* a loose bed of uncompacted powder), and will range up to about 8 for compacted powder bodies. Satterfield and Sherwood[¶] studied diffusion of hydrogen/air mixtures in a variety of media with $0.2 < \omega < 1$. They found that the ratio D_{eff}/D varies from 0.1 to 1. The majority of media they studied (which were all uncompacted powders) have tortuosity values close to 2.

There is a second mechanism which reduces gas diffusion through porous solids. Recall from Section 1.6 that the diffusion coefficient of a gas is related to its 'mean free path' λ. This is the average distance a gas molecule travels before it collides with another gas molecule. If the size of the pore channels become small enough, gas molecules will collide more often with the walls of the solid than with other molecules.

This process is known as Knudsen diffusion. Recall that the diffusion coefficient is proportional to λ (eq. (1.23)). Once Knudsen diffusion is dominant,

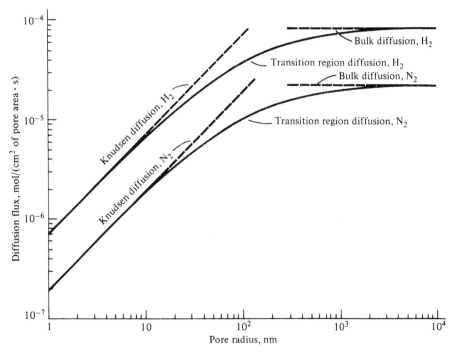

Figure 9.5 A plot of diffusive flux of hydrogen and nitrogen as a function of the pore radius, under 1 atmosphere total pressure at 298 K (C. N. Satterfield, *Heterogeneous Catalysis*, McGraw-Hill, New York, 1980, p. 338)

¶ C. N. Satterfield and T. K. Sherwood, *The Role of Diffusion in Catalysts* (Addison-Wesley, MA, 1963)

λ is replaced approximately by the pore channel diameter d. At elevated temperatures, λ is about 1 μm. Thus Knudsen diffusion is relevant to many microporous solids such as molecular filters. It is not important however for granular solids (*e.g.* sand), for which the pore channels are of millimetre dimensions. Figure 9.5 shows data for Knudsen diffusion in a hydrogen/nitrogen gas mixture as a function of the pore radius. The diffusivity is reduced for pores below about 1 μm and is dominated by the Knudsen effect once the pore radius reaches 30 nm or so.

9.4 Further reading

E. T. Turkdogan, *Physicochemical Properties of Molten Slags and Glasses* (Metals Society, London, 1983)

J. Szekely, J. W. Evans and H. Y. Sohu, *Gas–Solid Reactions* (Academic Press, New York, 1976)

D. R. Poirier and G. H. Geiger, *Transport Phenomena in Materials Processing* (TMS-AIME, Warrendale, PA, 1994)

Chemical Engineering Handbook, 5th edition, R. H. Perry and C. H. Chitton, eds. (McGraw-Hill, New York, 1973)

9.5 Problems to chapter 9

9.1   A heat transfer loop using liquid metal is used to cool a nuclear reactor. Liquid lead–bismuth eutectic alloy is pumped around the closed circuit, going from the reactor to the heat exchanger and back to the reactor. The reactor core heats the liquid metal to 500 °C. This heat is then removed in the cold leg of the loop using a heat exchanger system coupled to a steam turbine, and the temperature in the cold section of the loop is maintained at 375 °C. Most of the pipe work for pumping the lead–bismuth alloy is constructed of steel which has a small but significant solubility in the liquid alloy. This solubility increases with temperature according to

$$\log_{10}(\text{ppm Fe}) = 6.03 - \frac{4440}{T}$$

where ppm is parts per million on an atomic basis and T is in K. Since the bulk concentration of the iron leaving the hot zone is supersaturated with respect to the solubility limit of iron in the cold section, some mass transfer occurs, resulting in 'corrosion' of the piping in the hot leg section and deposition in the cold leg.

Obtain a relation between the bulk concentration C_0 of iron entering the hot leg section and the following variables:

L	length of hot zone	$0.3\,\text{m}$
ℓ	length of cold zone	$0.9\,\text{m}$
k	liquid-phase mass transfer coefficient of Fe in the alloy	$10^{-4}\,\text{m/s}$
u	bulk velocity in the liquid	$0.03\,\text{m/s}$
d	diameter of pipe	$0.08\,\text{m}$

Also, calculate the increase in internal diameter of the hot leg during two years of operation. You may assume that the dissolution is mass transfer controlled and that the alloy is in plug flow.

Pb–Bi eutectic composition: $X_{\text{Pb}} = 0.45,\ X_{\text{Bi}} = 0.55$

$\rho_{\text{Pb}} = 10.8\,\text{Mg/m}^3$ $\rho_{\text{Bi}} = 9.5\,\text{Mg/m}^3$

Atomic wt Pb $= 207.2\,\text{g/g\,mol}$ Atomic wt Bi $= 209.0\,\text{g/g\,mol}$

9.2   Calcium carbide has a high affinity for sulphur, and is therefore used as a desulphurizer. Consider the case of removal of S_2 from an argon stream for which the reaction can be represented as:

$$\text{CaC}_2 + \tfrac{1}{2}\text{S}_2 \rightarrow \text{CaS} + 2\text{C}$$

and the free energy of reaction is:

$$\Delta G^\circ = -1.05 \times 10^6 + 237.0T\,(\text{joules}).$$

Assume that we initially have a $0.05\,\text{m}$ diameter spherical particle of calcium carbide held in a stream of argon flowing at $10\,\text{m/s}$ containing 1% S_2 by volume, all at a temperature of $1473\,\text{K}$. We want to be able to calculate the rate of desulphurization from the gas stream.

It has been found by dissection that a topochemical product layer of CaS and C forms at the outer surface of the particles. Assume that sulphur gas is the diffusing species through this product layer, and that the void fraction and tortuosity in the product layer are 0.2 and 5, respectively. The sulphur diffusivity is $10^{-4}\,\text{m}^2/\text{s}$.

(a) Calculate the equilibrium constant at $1473\,\text{K}$ and the partial pressure and concentration of sulphur vapour at the reaction front.

(b) Sketch the schematic concentrations of the sulphur gas in the vicinity of the interface and in the product layer. Do this at the initial time and after various amounts of time. Clearly label your axes and the relative times. Put known concentrations on the diagram.

(c) What are the two possible rate-controlling steps in this process? Which is more likely to be the controlling step at long times?

(d) Set up the equation for the instantaneous flux of sulphur at some time of the order of an hour. (*Hint:* It takes much longer for the sphere to be completely converted to CaS.)

(e) Determine numerical values for the mass transfer coefficient and effective diffusivity in the product layer.

(f) Describe how your answer would be changed if the gas were:
(i) 50% S_2, 50% Ar, and
(ii) 100% S_2.

9.3   Less noble impurities in silver, such as copper and tin, can be eliminated by blowing pure oxygen onto the surface of molten silver. Silver oxide is unstable at such high temperatures. However, the oxides of copper and tin are solids which can be skimmed from the molten silver surface.

Consider, first of all, the case of no impurities in the silver. The solubility of oxygen in molten silver at $1000\,^\circ$C is $820\,\mathrm{mol/m^3}$.

(a) Indicate conditions in which the rate of oxygen dissolution in the silver is controlled by:
(i) diffusion in the gas phase,
(ii) diffusion in the liquid phase, and
(iii) the rate of oxygen blowing onto the bath surface.

(b) Calculate the time for the bath to become semi-saturated (concentration equal to one-half of the saturation value) with oxygen. Assume that the crucible containing the silver is a vertical cylinder with the data given below. Also assume that the rate of oxygen injection is high so that only a small fraction of the oxygen dissolves. Furthermore, assume that the injection creates convection in the bath, but does not cause waves on the surface.

(c) Now, consider the case of a dilute impurity such as copper in the silver. Assume that the vapour pressure of copper in this alloy is negligible. Draw a diagram of the concentrations of copper and oxygen on the gas and liquid sides in the vicinity of the gas/liquid interface. Do not calculate any numerical values for this part.

Data: Mass of silver   150 kg
        Overall surface mass transfer coefficient 5×10^{-4} m/s
        Depth of bath   0.25 m

9.4   Molten copper is deoxidized by the injection of natural gas (mainly methane, CH_4) into the copper. The reaction can be represented by:

$$CH_4 + 2\underline{O} \rightarrow C + 2H_2O.$$

The carbon forms soot which is an industrial hygiene problem, and the energy costs for this process are high. It has been proposed to use hydrogen for deoxidation to overcome these problems:

$H_2 + \underline{O} \rightarrow H_2O.$

It is your job to examine the fundamental aspects of this reaction.
(a) Sketch the concentration profiles for reactants and products at the bubble/copper interface.
(b) Assuming that equilibrium is established at the interface, the reaction can either be controlled by gas- or liquid-phase diffusion. For simplicity, just consider diffusion of oxygen in copper and water vapour in the bubble. Will gas-phase control be favoured by high or low values of the following parameter?
   (i) $\%\underline{O}$ in copper,
   (ii) mass transfer coefficient in the liquid, k_1
   (iii) mass transfer coefficient in the gas, k_g,
   (iv) equilibrium constant, K?
   State your reasons.
(c) Assuming that, for a particular situation, the rate is controlled by oxygen diffusion, calculate how much hydrogen will be left in a bubble by the time it leaves the copper, *i.e.* the hydrogen utilization. Assume that the bubble temperature remains constant at the copper temperature, 1200 °C, and that the total pressure is constant at 1 atmosphere (101.3 kPa).

For your calculations, you may use the following data:

$$K = \frac{p_{H_2O}}{p_{H_2}\%\underline{O}} = 2400 \ [(\%)]^{-1}$$

which can be converted to concentrations:

$$K = \frac{C_{H_2O}}{C_{H_2}C_{\underline{O}}} = 0.492 \ [\text{m}^3/\text{mol}].$$

Data: gas-phase mass transfer coefficient, k_g 0.1 m/s

liquid-phase mass transfer coefficient, k_1 3×10^{-4} m/s

bubble diameter, d_b 0.1 m

oxygen content in the copper   0.10% (496 mol/m³)

bubble velocity   0.5 m/s

depth of copper   1 m

(*Hint:* There is still considerable hydrogen left in the bubble at the bath surface, so that you can assume that the p_{H_2O}/p_{H_2} is less than 1 for your calculations.)

9.5   The removal of manganese from iron occurs by the transfer to the slag phase according to the reaction:

$$\underline{Mn} + FeO \rightarrow Fe + MnO,$$

where the oxides are dissolved in the slag. Assume chemical equilibrium at the slag/metal interface. Since iron is the major component in the metal phase, the transport of iron will be rapid. Also assume that there is a large concentration of FeO which does not change appreciably during manganese transport.
(a) Sketch the concentration profiles for Mn, Fe, MnO and FeO in the vicinity of the slag/metal interface.
(b) Develop an equation for the flux of manganese in terms of the volume of the metal, the equilibrium constant, the activity coefficients, the mass transfer coefficients and the bulk concentrations.
(c) In practice, the slowest step in the process is the manganese transfer from metal phase to the reaction interface. The following information is available:
(1) the metal bath depth is 0.3 m and slag depth is 0.1 m;
(2) the mass transfer coefficient of manganese on the metal side is 3×10^{-4} m/s.
Calculate the time to reach 90% of the difference from the starting manganese concentration to the equilibrium concentration.

9.6   Spherical particles of zinc blende, ZnS, of radius $r = 1$ mm are roasted in an 8% oxygen stream at 900 °C and 1 atm total pressure. The stoichiometry of the reaction is $2ZnS + 3O_2 \rightarrow 2ZnO + 2SO_2$. Assuming that the reaction proceeds by a topochemical reaction model:
(a) Calculate:
(i) the time needed for complete conversion of a particle;
(ii) the overall resistance of the product layer during this operation.
(b) Repeat for particles of radius $r = 0.05$ mm.
Data: density, $\rho_{solid} = 4.13$ Mg/m^3
reaction rate constant, $k = 0.02$ m/s
for gases in the ZnO layer, $D = 8 \times 10^{-6}$ m^2/s
Note that film resistance can be neglected as long as a growing product layer is present.

9.7   Solid spherical particles of differing size are introduced into a constant temperature furnace and kept there for one hour. Under these conditions, the 4 mm diameter particles are 58% converted, the 2 mm diameter particles are 87.5% converted.
(a) Find the rate-controlling mechanism for the conversion of solids.

    (b) Find the time needed for complete conversion of 1 mm diameter particles in this furnace.

9.8   Sulphur dioxide is absorbed by water from the mixture of sulphur dioxide and air. The concentration of SO_2 in the gas inside the absorption tower is 10 vol.% and that in the liquid which is in contact with this gas is 0.4 wt%. Density and temperature of the liquid are 0.991 and 50 °C, respectively. The total pressure is 1 atm. The overall mass transfer coefficient based on the gas concentration is $k_{ov} = 268$ g mol $SO_2/hr\,m^2$ atm. 47% of the multi-phase resistance is due to mass transfer in the gas phase and 53% is due to the liquid phase. Equilibrium data at 50 °C are as follows:

SO_2 concentration in liquid (g/100 g)	0.2	0.3	0.5	0.7
SO_2 partial pressure (mm Hg)	29	46	83	119

    (a) Calculate the overall mass transfer coefficient based on the concentration in the liquid phase.

    (b) Calculate the composition at the interface between both phases.

Part D

Alternative driving forces for diffusion

Chapter 10

General driving force for diffusion

Applied science is a conjuror, whose bottomless hat yields
impartially the softest of Angora rabbits and the most
petrifying of Medusas.
Aldous Huxley, *Tomorrow and Tomorrow and Tomorrow*

*Up to now we have assumed that only concentration gradients provide a driving force
for diffusion. We now generalize this, by including all forms of free-energy gradient.
This will lead us fairly easily to a general theory of diffusion that includes Fick's First
Law as a special case. We will apply our new understanding to problems such as
diffusion due to an electrical field or a mechanical stress. We will finish the chapter by
seeing how this method can help to understand diffusion in multicomponent systems.*

10.1 **Theory**

Throughout this book we have considered that diffusional processes are solely
driven by concentration gradients. This is based on the concept that a material
strives towards equilibrium, and that equilibrium is defined as a state of uniform concentration. In materials science, however, we recognize that this is a
rather naive definition of equilibrium. To be more precise, we should state that
a body is at equilibrium when the free energy is uniform throughout. Using
this definition, we can broaden considerably the range of situations in which
diffusion occurs. Thus, we expect diffusion not only in response to concentration gradients, but also due to electrical and thermal fields, and mechanical

stress. We are also better able to understand the complex diffusion phenomena which occur in multi-component alloys.

You will recall that, in Chapter 1, we developed a simple picture of diffusion at the atomic scale. This was based on the idea that atoms in a lattice occupy local minima in the free energy. We assumed that this energy was uniform throughout the solid (see Fig. 1.3). In order for an atom (or vacancy) to move to an adjacent site, it had to overcome an activation barrier, equal to ΔG^m.

Now let us suppose that the free energy of atoms, in their rest positions, is not uniform everywhere, but is as pictured in Fig. 10.1. We can envisage this gradient in free energy as a force given by

$$F = -\frac{\partial G}{\partial y} = -\frac{\Delta G}{a}, \tag{10.1}$$

where $a \, (= y_2 - y_1)$ is the distance between neighbouring minima. We noted in Chapter 1 that atoms jump between adjacent positions at a frequency $\Gamma = \nu p$, where ν is the vibration frequency, and p is the probability that the atom has sufficient energy to make the jump. We further noted that $p = \exp(-\Delta G^m/kT)$. From this we came up with the diffusive flux in one dimension, eq. (1.6),

$$J = g\Gamma a(C_1 - C_2).$$

Note that we have expressed this in terms of the more general geometric parameter g rather than use the specific value $\frac{1}{2}$ given in eq. (1.6), which is strictly valid only for a one-dimensional solid.

The present case is a bit more complex, since the height of the activation barrier is different for jumps from plane 1 to plane 2, than for jumps from plane 2 to plane 1. We therefore re-write eq. (1.6) as

$$J = ga(\Gamma_{12}C_1 - \Gamma_{21}C_2), \tag{10.2}$$

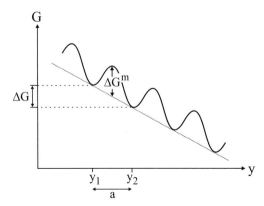

Figure 10.1 Free energy vs position on an atomic scale when there is a net driving force in the positive y-direction.

where

$$\Gamma_{12} = \nu x \exp\left[-\frac{\left(\Delta G^{\mathrm{m}} - \frac{1}{2}\Delta G\right)}{kT}\right] \tag{10.3a}$$

and

$$\Gamma_{21} = \nu x \exp\left[-\frac{\left(\Delta G^{\mathrm{m}} + \frac{1}{2}\Delta G\right)}{kT}\right]. \tag{10.3b}$$

Note that x represents the probability that an adjacent site is vacant. Thus, $x \approx 1$ for interstitial or vacancy diffusion (see eq. (1.8) and (1.11), respectively) while $x = X_{\mathrm{v}}$, the vacancy concentration, for self-diffusion and the diffusion of substitutional solutes (eq. 1.12). Combining eqs. (10.2) and (10.3) gives the general result:

$$J = ga\nu x \exp\left(-\frac{\Delta G^{\mathrm{m}}}{kT}\right)\left[C_1 \exp\left(\frac{\Delta G}{2kT}\right) - C_2 \exp\left(\frac{-\Delta G}{2kT}\right)\right]. \tag{10.4}$$

This is a rather nasty-looking equation. We are however saved by the fact that for *almost* all cases of practical interest $\Delta G \ll kT$.[¶] We note that $e^x \approx 1 + x$ when $x \ll 1$. Therefore, eq. (10.4) simplifies to

$$J = ga\nu x \exp\left(-\frac{\Delta G^{\mathrm{m}}}{kT}\right)\left[C_1\left(1 + \frac{\Delta G}{2kT}\right) - C_2\left(1 - \frac{\Delta G}{2kT}\right)\right]. \tag{10.5}$$

You will see that the term outside the square brackets equals D/a (see, for example, eqs. (1.11) and (1.13)). Thus we end up with a rather simple result.

$$J = \frac{D}{a}\left[(C_1 - C_2) + (C_1 + C_2)\frac{\Delta G}{2kT}\right]$$

or, using $F = -\Delta G/a$ from eq. (10.1), and setting $C = \frac{1}{2}(C_1 + C_2)$ as the average concentration,

$$J = -D\frac{\partial C}{\partial y} + \frac{DC}{kT}F. \tag{10.6}$$

We therefore find that so long as $\Delta G \ll kT$, the diffusive flux can be written as the sum of two simple components. The first is, of course, just Fick's First Law and represents the response of the system to a concentration gradient. The second term represents the additional response due to a force or chemical potential gradient in the material. We call this second term the Einstein drift equation.

[¶] The exception occurs near singularities such as in dislocation cores and at crack tips. Here the free-energy gradients can be very large indeed.

10.2   **Diffusion due to an electrical field**

Suppose that you apply an electrical field to an ionic solid. This field exerts a force on the ions, with positive and negative ions experiencing a force in opposite directions. The magnitude is given by

$$F = q\mathscr{E} = Ze\mathscr{E}, \tag{10.7}$$

where $\mathscr{E}$ is the electrical field and q is the charge on the ion given as the product of the valence Z and the electron charge e. If the concentration of the ion is uniform throughout the solid, the only driving force for diffusion is that given by the electrical field. Thus, only the second term in eq. (10.6) is applicable.

Now we know that when charged particles flow in response to an electrical field this produces a current. We are used to thinking about this in terms of electrons flowing in a metal. However, the result is similar when ions move in a solid. It should therefore be possible to calculate the electrical conductivity and relate it to the diffusivity. To start, we recall that, from Ohm's Law, $I = \sigma\mathscr{E}$, where I is electrical current density and σ is electrical conductivity. Since the current density results from a flux of ions we can write

$$I = JZe. \tag{10.8}$$

We can now substitute these relations into the Einstein Drift equation, as follows:

$$\sigma\mathscr{E} = I = JZe = \frac{CD}{kT} ZeF = \frac{CD}{kT} (Ze)^2 \mathscr{E}. \tag{10.9}$$

Thus, the electrical conductivity and diffusivity are directly related as

$$\sigma = \frac{CD}{kT} (Ze)^2. \tag{10.10}$$

This is known as the Nernst–Einstein equation, and has been confirmed by independent electrical and diffusion measurements in materials undergoing ionic conduction.

In pure, stoichiometric compounds, the diffusion coefficient, and thus the ionic conductivity is low. However, for more complex compounds with large defect concentrations, ionic conduction can be quite rapid. For example, CaO-doped ZrO_2 exhibits a conductivity of about 0.01/ohm m at 500 °C. This is an order of magnitude higher than the conductivity of Si at room temperature. The addition of Ca in the ZrO_2 produces excess O^{2-} vacancies. Thus the high conductivity in this system is due to oxygen-ion conduction. This means that ZrO_2 ceramics can be used in high-temperature probes to measure the oxygen content in, for example, molten metal or high-temperature engines. The oxygen diffuses in response to an oxygen differential from one side of the probe to the other.

Another example of an important ion conductor is sodium aluminate, commonly known as β-alumina. This material has a layered structure containing a poorly populated plane of Na^+ ions. These ions can therefore diffuse rapidly through the structure. Such materials can be used as solid electrolytes in fuel cells. An electrolyte is a material which mechanically isolates two sides of a cell but which enables transport of one of the reacting species. One such system is based on liquid Na and solid S as the reactants with a β-alumina electrolyte. Such a fuel cell would be four times lighter and half the size of a conventional Pb/PbO_2 battery.

10.3   Diffusion due to mechanical stress

10.3.1   Vacancy diffusion due to local forces

When a load is applied to a body there is a driving force to change its shape. Thus, under a tensile load, the body will want to become longer in the direction of the load. At elevated temperatures, diffusion will contribute to this, by a process known as diffusion creep. Here, we will examine this process for a pure, single-component material. The same behaviour occurs in alloys and compounds, although the details are more complex.

As we know, diffusion in a pure material occurs through a vacancy process. Consider a crystal, as illustrated in Fig. 10.2, which is subjected to a tensile stress. In any solid at high temperatures, point defects (such as vacancies) are constantly created and annihilated at free surfaces. The creation of vacancies involves a process such as that illustrated in Fig. 10.3(a). The rate at which

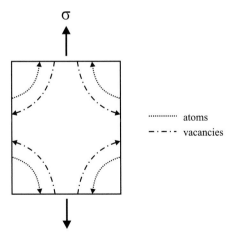

Figure 10.2 Vacancy and atom fluxes in a single crystal subjected to a tensile stress.

such a process occurs on a surface is equal to the number of suitable sites (such as ledges) n_s, times the probability that the atom has sufficient energy to make this jump. Thus the vacancy creation rate $\dot{n}_c$ is

$$\dot{n}_c = n_s \nu \exp\left(-\frac{\Delta G_v^f + \Delta G_v^m}{kT}\right) = n_s \nu \exp\left(-\frac{\Delta G_v^c}{kT}\right). \tag{10.11}$$

For this process, the free energy of vacancy creation, ΔG_v^c, contains two terms, that for vacancy migration, and that for the formation of vacancies.

A similar process occurs when a vacancy is annihilated at a surface (see Fig. 10.3(b)), except that there is no energy of formation. Thus

$$\dot{n}_a = n_s \nu X_v \exp\left(-\frac{\Delta G_v^m}{kT}\right). \tag{10.12}$$

Note, however, that for this case we must include the probability that a vacancy is adjacent to a suitable site. This is the vacancy concentration X_v. At equilibrium, $\dot{n}_a = \dot{n}_c$, and therefore as we expect (see eq. (1.3))

$$X_v = X_v^o = \exp\left(-\frac{\Delta G_v^f}{kT}\right).$$

Now, if a mechanical stress σ is applied to the surface, additional work ΔW is involved in vacancy creation. Thus the energy for vacancy creation has the form

$$\Delta G_v^c = \Delta G_v^f + \Delta G_v^m - \Delta W. \tag{10.13}$$

The work term is negative, since a positive (tensile) stress enhances vacancy formation (*i.e.* it lowers ΔG_v^c). As a result of this, the vacancy concentration

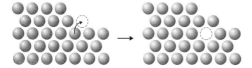

(a) Vacancy creation

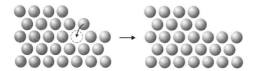

(b) Vacancy annihilation

Figure 10.3 A schematic illustration of (a) vacancy creation and (b) vacancy annihilation at a free surface.

just below a surface under stress is biased, and has an equilibrium value given by

$$X_v^b = \exp\left[-\frac{(\Delta G_v^f - \Delta W)}{kT}\right]$$

$$= X_v^o \exp\left(\frac{\Delta W}{kT}\right). \tag{10.14}$$

In order to determine how ΔW is related to stress, we perform a simple virtual work calculation. Consider Fig. 10.4, which depicts the diffusion in a crystal subjected to a tensile stress. Now let us suppose we add a new layer of atoms to the top surface (of area A) through the creation of vacancies which diffuse into the crystal. The work done by the stress (force times displacement) is equal to $(\sigma A)a$ per atomic layer, where a is the lattice parameter. The number of atoms per layer is Aa/Ω, where Ω is the atomic volume. Thus the work done per atom is

$$\Delta W = \sigma\Omega. \tag{10.15}$$

As a result, the vacancy concentration on a surface subjected to stress is given by

$$X_v^b = X_v^o \exp\left(\frac{\sigma\Omega}{kT}\right). \tag{10.16}$$

However, since the work done is generally much less than the thermal energy $(\sigma\Omega \ll kT)$, we can linearize this result to

$$X_v^b = X_v^o \left(1 + \frac{\sigma\Omega}{kT}\right). \tag{10.17}$$

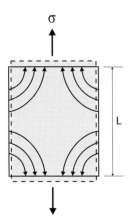

Figure 10.4 A schematic illustration of the diffusion process leading to crystal elongation by creep.

10.3.2 Application to diffusion creep

When a tensile stress is applied to a crystal as shown in Fig. 10.4, the vacancy concentration is increased at the top and bottom surface according to eq. (10.17). However, on the side surfaces, where the normal stress is zero, there is no enhanced vacancy concentration and $X_v = X_v^o$. Thus the free-energy difference for vacancies between the vertical and horizontal surfaces is $\sigma\Omega$. The separation between the mid-point of these surfaces is $L/\sqrt{2}$. Therefore, the driving force for vacancy diffusion is $\sqrt{2}\sigma\Omega/L$. We can substitute this into the Einstein Drift equation to determine the vacancy flux

$$J_v = \sqrt{2}\,\frac{C_v^o D_v \sigma\Omega}{kTL} = \sqrt{2}\,\frac{X_v^o D_v \sigma}{kTL} \tag{10.18}$$

since $X_v^o = C_v^o\Omega$.

Now we are really interested in the diffusion of atoms, not vacancies. Since $J_a = J_v$ (but in the opposite direction), and $D_a = D_v X_v^o$, then

$$J_a = \sqrt{2}\,\frac{D_a \sigma}{kTL}. \tag{10.19}$$

While we have developed this result for a single crystal, it applies equally well for polycrystals, just by replacing the crystal size L by the grain size d. This is because internal surfaces, such as grain boundaries and incoherent interfaces, are just as effective as sites for vacancy creation and annihilation as free surfaces. This process is known as Nabarro–Herring creep.

When we perform creep experiments, we do not measure diffusion flux directly. Instead, we monitor the strain rate $\dot{\varepsilon}$ as a function of the applied tensile stress σ.[¶] Through simple conservation arguments we can show that $\dot{\varepsilon} = J\Omega/d$ and thus

$$\dot{\varepsilon} = \sqrt{2}\,\frac{D_a \sigma\Omega}{kTd^2}. \tag{10.20}$$

This is the Nabarro–Herring creep equation (although in a more precise model the numerical terms are altered slightly).[†] By a similar model, it is possible to derive an equation for the case in which diffusion proceeds along grain boundaries (Coble creep):[‡]

$$\dot{\varepsilon} = \alpha\,\frac{\delta_b D_b \sigma\Omega}{kTd^3}, \tag{10.21}$$

where α is a numerical constant.

[¶] Strain is defined as the fractional change of length during a deformation process.
[†] F. R. N. Nabarro, *Strength of solids* (The Physical Society, London, 1948), p. 75; C. Herring, *J. Appl. Phys.*, **21**, 37 (1950)
[‡] R. L. Coble, *J. Appl. Phys.*, **34**, 1679 (1963)

10.4   Diffusion due to surface curvature

An additional driving force for diffusion is surface curvature. This is respon-
sible for processes such as solid-state sintering and Ostwald ripening. For any
body of finite dimensions the total free energy includes a term due to the excess
free energy at surfaces. For large bodies, this term is negligible. However,
when the ratio of surface area to volume increases (as in small particles or
along rough surfaces) this is no longer the case. The excess free energy at a
curved surface is $\Delta W = \gamma_s \Omega K$, where γ_s is the surface free energy and K is
the curvature ($K = 2/r$ for a sphere of radius r). We can use all the results of
the previous section, with a new definition of ΔW to ascertain the effect of
curvature on the diffusive flux.

Consider the case of solid-state sintering of two spherical particles (see
Fig. 10.5). There are three points of interest:

(a)   on the surface, away from the neck, $\Delta W = -\dfrac{2\gamma_s \Omega}{r}$,

(b)   on the surface, at the neck, $\Delta W = \gamma_s \Omega \left(\dfrac{1}{\rho} - \dfrac{1}{d} \right) \approx \dfrac{\gamma_s \Omega}{\rho}$, where ρ is the

      radius of curvature in the neck between the particles (see Fig. 10.5),

(c)   in the centre of the neck, $\Delta W \approx 0$.

This sets up two separate driving forces for vacancy migration. One occurs
between regions of high and low curvature along the surface of the sphere
(which results in neck growth without densification, paths ②, ③ and ④ in
Fig. 10.5). The other [force] arises due to the potential gradient between the
centre of the neck and the neck surface (which results in neck growth with

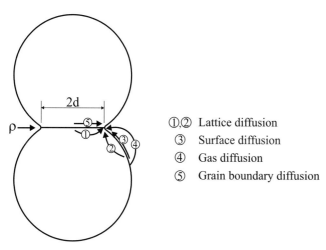

①,②	Lattice diffusion
③	Surface diffusion
④	Gas diffusion
⑤	Grain boundary diffusion

Figure 10.5 An illustration of diffusion during sintering driven by surface-curvature
gradients.

densification, paths ① and ⑤). A more complete description of these processes and how they are modelled is given by Ashby (1974).[¶]

10.5   Diffusion due to an activity gradient

All of the various driving forces for diffusion represent differences in the free energy of atoms from one position in a body to another. The molar free energy of any species A can be written as

$$\overline{G}_A = G_A^o + RT \ln a_A \tag{10.22}$$

where G_A^o is the free energy of A in its standard state and a_A is the thermo-dynamic activity of A in the body of interest. We can also write this in terms of the free energy per atom by dividing by Avogadro's number N_0:

$$(G_A)_{\text{atom}} = \frac{G_A^o}{N_0} + kT \ln a_A. \tag{10.23}$$

When the concentration gradient is the only driving force for diffusion we use Fick's First Law which says, in essence, that the flux is proportional to the concentration gradient. By analogy, we can generalize this by recognizing that the diffusion acts to remove free-energy gradients. We can therefore write that

$$J_A = -L_A \frac{\partial (G_A)_{\text{atom}}}{\partial y} = -kTL_A \frac{\partial \ln a_A}{\partial y}. \tag{10.24}$$

Note that we have used a new mobility constant L_A. This must be related to the diffusion coefficient but may not be exactly the same.

  We can determine L_A if we realize that for the case in which a concentration gradient is the only driving force, eq. (10.24) must become Fick's First Law. So recall that the activity

$$a_A = \gamma_A \cdot X_A, \tag{10.25}$$

where γ_A is the activity coefficient. When concentration is the only driving force then γ_A must be constant, and

$$\frac{\partial \ln a_A}{\partial y} = \frac{\partial \ln X_A}{\partial y} = \frac{1}{X_A}\frac{\partial X_A}{\partial y} = \frac{1}{C_A}\frac{\partial C_A}{\partial y}. \tag{10.26}$$

If we substitute this into eq. (10.24), we find that

$$J_A = -\frac{kTL_A}{C_A}\frac{\partial C_A}{\partial y} = -D_A \frac{\partial C_A}{\partial y}. \tag{10.27}$$

[¶] M. F. Ashby, *Acta Metall.*, **22**, 275 (1974)

Thus, $kTL_A = D_A C_A$. This relationship must apply generally, so that

$$J_A = -C_A D_A \frac{\partial \ln a_A}{\partial y} \qquad (10.28)$$

gives the flux due to an activity gradient.

10.5.1 Application to 'uphill' diffusion

We will now apply this result to the rather curious phenomenon of 'uphill' diffusion, in which atoms diffuse *up* a concentration gradient. We will refer to a classic experiment presented by Darken.[¶] He made a diffusion couple, in which one side was an Fe–0.44% C alloy and the other an Fe–0.48% C–3.8% Si alloy. These were annealed at 1050 °C for several days. At this temperature, both alloys are austenitic so only a single phase exists. According to Fick's First Law, carbon should diffuse towards the Si-free alloy in order to remove the small difference in concentration between the two halves of the couple. However, after 13 days at 1050 °C, the carbon concentration indicated that a rather different diffusion process had taken place (as shown in the profile given in Fig. 10.6).

Now, it is important to realize that C, being interstitial, has a much higher diffusion coefficient than does Si. Thus, very little Si diffusion has occurred – it is essentially still in the left side of the couple. But why has the carbon moved towards the alloy with the high C concentration? The reason for this behaviour is that the carbon *activity* a_c in Fe is increased by the presence of Si. This is shown schematically in Fig. 10.7. In Fig. 10.7(a), we see an activity profile. Once the couple is heated to 1050 °C, local equilibrium will be quickly established (*i.e.* the activities will be equal at the interface). Carbon diffusion will occur down the activity gradient. The solution is identical to that which we

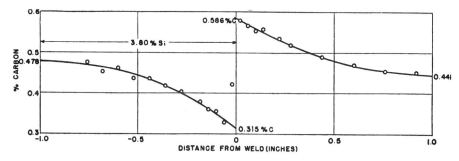

Figure 10.6 The carbon concentration profile measured after 13 days at 1050°C (Darken, 1949). The left side was originally Fe–0.48% C–3.8% Si, while the right side was Fe–0.44%C.

[¶] L. Darken, *Trans. AIME,* **180**, 430 (1949)

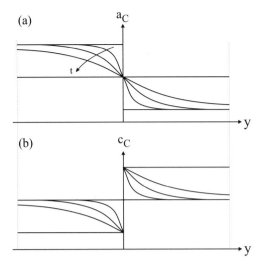

Figure 10.7 (a) A schematic illustration of the activity profiles across Darken's diffusion couple at different times, before substantial Si diffusion has occurred. (b) The corresponding concentration profiles.

developed in Section 3.3, except that we use ln (activity) rather than concentration. If we convert the activity curves in Fig. 10.7(a) to concentrations, the result is Fig. 10.7(b). The effects at the interface are due to the abrupt change of activity coefficient. Thus carbon appears to diffuse 'uphill'.

Finally, you should realize that if we wait long enough the Si will diffuse across this couple, altering the activity coefficient of C as it does so. The final equilibrium state will therefore be reached once the whole couple is of uniform concentration.

10.6   Further reading

J. S. Kirkaldy and D. J. Young, *Diffusion in the Condensed State* (Institute of Metals, London, 1987)

P. Shewmon, *Diffusion in Solids* 2nd edition (TMS-AIME, Warrendale, PA, 1989)

B. Burton, *Diffusional Creep of Polycrystalline Materials* (Trans Tech, Aedermannsdorf, Switzerland, 1977)

10.7   Problems to chapter 10

10.1   (a) Starting from first principles, derive the Einstein drift equation which describes the diffusive flux J due to an applied drive force F.

(b) Use this result to show that, in an ionic solid which conducts electricity purely by ionic diffusion, the conductivity σ is related to the diffusivity D, by a relation of the form,

$$\sigma = (Ze)^2 \frac{CD}{kT},$$

the Nernst-Einstein equation.

10.2   (a) Starting from first principles, derive the Einstein drift equation (in one dimension). Be sure to state explicitly any assumptions you make, and why.

(b) The compound LiAl is a conductor of Li^+ ions. It has this property because of a high, thermodynamically stable concentration of Li vacancies, equal to 1%. Derive an expression for the diffusion coefficient of Li^+ ions, and determine a numerical value at room temperature.

(c) Use this value and the Einstein drift equation developed above, to determine the conductivity of Li^+ ions in LiAl, at room temperature (*i.e.* find a numerical value).

10.3   CoO is a p-type oxide with an excess of vacancies on Co sites (V_{Co}). Given this, and the fact that the activation energy for the diffusion of Co in CoO is about half of that for O diffusion, which element do you expect will control the rate of scale growth in CoO? Explain your reasoning.

Appendix A

Mathematical methods for the solution of Fick's Second Law

A.1   Separation of variables method

We will assume that the solution for solute concentration $C(y, t)$ can be separated into two functions, one depending only on position, the other on time, *i.e.*

$$C(y, t) = \chi(y)\Theta(t),$$

If we substitute this into Fick's Second Law then

$$\chi \frac{d\Theta}{dt} = D\Theta \frac{d^2\chi}{dy^2}.$$

We then separate the y- and t-dependent terms,

$$\frac{1}{D\Theta} \frac{d\Theta}{dt} = \frac{1}{\chi} \frac{d^2\chi}{dy^2} = -\lambda^2. \tag{A.1}$$

For the first equality in this expression to be valid, each side must be constant. We set this constant to be $(-\lambda^2)$. We now solve each side of the equation separately. The time-dependent equation

$$\frac{1}{\Theta} \frac{d\Theta}{dt} = -D\lambda^2$$

can be integrated to yield

$$\Theta(t) = \Theta_0 \exp(-\lambda^2 Dt), \tag{A.2}$$

where Θ_0 is a constant of integration. The spatial equation

$$\frac{1}{\chi}\frac{\mathrm{d}^2\chi}{\mathrm{d}y^2} = -\lambda^2$$

is integrated twice, giving

$$\chi(y) = A'\sin(\lambda y) + B'\cos(\lambda y), \tag{A.3}$$

where A' and B' are integration constants. Combining these expressions gives

$$C(y,t) = [A\sin(\lambda y) + B\cos(\lambda y)]\exp(-\lambda^2 Dt), \tag{A.4}$$

where $A = A'\Theta_0$ and $B = B'\Theta_0$. In general, there are a multitude of solutions to Fick's Second Law, all of which have this form, but with different values of A, B and λ. Therefore, the most general solution is given by the sum of these

$$C(y,t) = \sum_{n=0}^{\infty}[A_n\sin(\lambda_n y) + B_n\cos(\lambda_n y)]\exp(-\lambda_n^2 Dt). \tag{A.5}$$

This solution is generally valid for diffusion in one dimension. Similar expressions can be derived for cylindrical or spherical symmetry.

In order to proceed further we need to define the initial and boundary conditions appropriate to the problems at hand. We will consider only one example, namely diffusion out of a plate with an initially uniform concentration, and the concentration fixed at the surface of the plate. The initial and boundary conditions are therefore given by

$$
\begin{array}{llll}
\text{I.C.} & C = C_i, & -L < y < L, & t = 0 \\
\text{B.C.} & C = C_s, & y = L & \\
 & \dfrac{\partial C}{\partial y} = 0, & y = 0. &
\end{array}
$$

Here L is the half-thickness of the plate. We have used the symmetry conditions at $y = 0$ as one boundary condition. (We could just as easily have exchanged this for the condition $C = C_s$ at $y = -L$). In order to simplify the analysis we make a simple transformation by setting

$$\xi = C - C_s.$$

Since

$$\frac{\partial \xi}{\partial t} = \frac{\partial(C - C_s)}{\partial t} = \frac{\partial C}{\partial t}$$

and

$$\frac{\partial^2 \xi}{\partial y^2} = \frac{\partial^2 (C - C_s)}{\partial y^2} = \frac{\partial^2 C}{\partial y^2}$$

this transformation does not affect Fick's Second Law. Therefore, the general solution just developed is still valid. The boundary values become

I.C.	$\xi = C_i - C_s,$	$-L < y < L, \quad t = 0$	(A.6)
B.C.	$\xi = 0,$	$y = L$	(A.7)
	$\dfrac{\partial \xi}{dy} = 0,$	$y = 0.$	(A.8)

We are now ready to solve the general equation, eq. (A.5) subject to these conditions. We first substitute eq. (A.8) into eq. (A.5). The result is

$$\frac{\partial \xi}{\partial y} = 0 = \sum_{n=0}^{\infty} [A_n \lambda_n \cos(\lambda_n y) - B_n \lambda_n \sin(\lambda_n y)] \exp(-\lambda_n^2 Dt)$$

at $y = 0$. This equality is only satisfied generally if all of the A_ns are equal to zero. Therefore,

$$\xi(y, t) = \sum_{n=0}^{\infty} B_n \cos(\lambda_n y) \exp(-\lambda_n^2 Dt). \tag{A.9}$$

We now substitute eq. (A.7) into this, *i.e.* $\xi = 0$ at $y = L$. This can be achieved if all the B_ns equal zero. But this results in a trivial solution. Therefore, the $\cos(\lambda_n L)$ terms must all equal zero. We know that $\cos(\lambda_n L) = 0$ only if

$$\lambda_n L = \frac{\pi}{2}, \frac{3\pi}{2}, \text{etc.}$$

In other words

$$\lambda_n L = \frac{2n - 1}{2} \pi. \tag{A.10}$$

Upon substitution back into eq. (A.9) this gives

$$\xi(y, t) = \sum_{n=0}^{\infty} B_n \cos\left(\frac{2n + 1}{2} \pi \frac{y}{L}\right) \exp\left[-\left(\frac{2n + 1}{2} \frac{\pi}{L}\right)^2 Dt\right].$$

The B_n values are now determined by substituting the initial conditions, eq. (A.6), into this relationship. Thus

$$\xi = \sum_{n=0}^{\infty} B_n \cos\left(\frac{2n + 1}{2} \pi \frac{y}{L}\right) = C_i - C_s \text{ at } t = 0. \tag{A.11}$$

The solution to this equation requires a Fourier analysis. In this technique we multiply both sides of the last expression by $\cos(\lambda_m y)$ and integrate between $-L$ and $+L$.

Thus

$$(C_i - C_s) \int_{-L}^{+L} \cos(\lambda_m y) dy = \sum_{n=0}^{\infty} B_n \int_{-L}^{+L} \cos(\lambda_n y) \cos(\lambda_m y) \, dy.$$

But it is known that

$$\int_{-L}^{+L} \cos(\lambda_m y) \, dy = (-1)^{m+1} \frac{4}{2m+1} \frac{L}{\pi}.$$

Moreover,

$$\int_{-L}^{+L} \cos(\lambda_n y)(\lambda_m y) \, dy = \begin{cases} 0 & \text{if } m \neq n \\ L & \text{if } m = n. \end{cases}$$

Therefore only one term is left and

$$B_m = \frac{4}{2m+1} \frac{(-1)^{m+1}}{\pi} (C_i - C_s). \tag{A.12}$$

Substituting this back into eq. (A.11) gives the final solution:

$$\frac{C(y,t) - C_s}{C_i - C_s} = \frac{4}{\pi} \sum_{n=0}^{\infty} \frac{(-1)^{n+1}}{2n+1}$$

$$\cdot \cos\left(\frac{2n+1}{2} \pi \frac{y}{L}\right) \exp\left[-\left(\frac{2n+1}{2} \frac{\pi}{L}\right)^2 Dt\right]. \tag{A.13}$$

The average concentration in the plate can also be determined. This is found by integrating over the plate thickness, *i.e.*

$$\bar{C} = \frac{1}{2L} \int_{-L}^{+L} C(y,t) \, dy = \frac{1}{L} \int_{0}^{+L} C(y,t) \, dy. \tag{A.14}$$

Substituting eq. (A.13) gives

$$\frac{\bar{C} - C_s}{C_i - C_s} = \frac{1}{L} \frac{4}{\pi} \sum_{n=0}^{\infty} \frac{(-1)^{n+1}}{2n+1} \exp\left[-\left(\frac{2n+1}{2} \frac{\pi}{L}\right)^2 Dt\right]$$

$$\cdot \int_{0}^{L} \cos\left(\frac{2n+1}{2} \pi \frac{y}{L}\right) dy$$

$$= \frac{8}{\pi^2} \sum_{n=0}^{\infty} \frac{1}{(2n+1)^2} \exp\left[-\left(\frac{2n+1}{2} \frac{\pi}{L}\right)^2 Dt\right]. \tag{A.15}$$

It is worth noting that each term in eqs. (A.13) and (A.15) is smaller than the previous one and the solution involves terms of alternating sign. Therefore the error involved in stopping after n terms is of the same order as the $(n+1)$th term. Moreover, the segregation between the terms increases with time. For example, the ratio between the first and third term is $3 \exp[2\pi^2 Dt/L^2]$. This ratio exceeds 100 if $Dt \geq 0.2L^2$, *i.e.* only a 1% error results from using the first term of the solution once this condition is met.

A.2   Laplace transform method

The Laplace transform approach is based physically on the plane initial source. It is therefore not surprising that the result takes the form of error functions. A Laplace transform is a device to transfer a partial differential equation (PDE) into an ordinary differential equation (ODE). Once the transformed equation is solved, the result must undergo a reverse transformation. We define the Laplace transform as follows

$$\bar{f}(p) = \int_0^\infty \mathrm{e}^{-pt} f(t)\, \mathrm{d}t.$$

We first consider the individual elements of Fick's Second Law. If $f(t) = \partial^2 C/\partial y^2$, then

$$\bar{f}(p) = \int_0^\infty \mathrm{e}^{-pt} \frac{\partial^2 C}{\partial y^2}\, \mathrm{d}t = \frac{\partial^2}{\partial y^2} \int_0^\infty C \mathrm{e}^{-pt}\, \mathrm{d}t = \frac{\partial^2 \bar{C}}{\partial y^2}, \qquad (A.16)$$

where $\bar{C}$ is the Laplace transform of C. If $f(t) = \partial C/\partial t$, however,

$$\bar{f}(p) = \int_0^\infty \mathrm{e}^{-pt} \frac{\partial C}{\partial t}\, \mathrm{d}t.$$

Integration by parts leads to

$$\bar{f}(p) = p\bar{C}. \qquad (A.17)$$

Therefore, the transformed version of Fick's Second Law is

$$\frac{\mathrm{d}^2 \bar{C}}{\mathrm{d}y^2} = p\bar{C}. \qquad (A.18)$$

Since $\bar{C}$ is a function of y only, this is an ordinary differential equation.

We also need to transform the boundary conditions. For example, if $C = C_\mathrm{s}$, then $\bar{C} = C_\mathrm{s}/p$. Similarly, the transform of $\partial C/\partial y = 0$ is $\mathrm{d}\bar{C}/\mathrm{d}y = 0$. Therefore, the case of a plate with a fixed surface concentration $C = C_\mathrm{s}$ at $y = \pm L$, can be treated by applying the transformed condition $\bar{C} = C_\mathrm{s}/p$ at $y = \pm L$ to eq. (A.18). The solution to this is

$$\bar{C} = C_\mathrm{s} \frac{\cosh(qy)}{p\cosh(qL)} \qquad (A.19)$$

where $q^2 = p/D$. This function has no inverse transform. However, it can be expanded as

$$\bar{C} = \frac{C_s}{p}\{\exp[-q(L-y)] + \exp[-q(L+y)]\}$$

$$\cdot \sum_{n=0}^{\infty}(-1)^n \exp(-2nqL)$$

$$= \frac{C_s}{p}\sum_{n=0}^{\infty}(-1)^n \exp\{-q[(2n+1)L - y]\}$$

$$+ \frac{C_s}{p}\sum_{n=0}^{\infty}(-1)^n \exp\{-q[(2n+1)L + y]\}. \qquad (A.20)$$

The exponential function does have an inverse transform. It is the error function. Thus, the solution becomes

$$C = C_s \sum_{n=0}^{\infty}(-1)^n \operatorname{erfc}\left[\frac{(2n+1)L - y}{2\sqrt{Dt}}\right]$$

$$+ C_s \sum_{n=0}^{\infty}(-1)^n \operatorname{erfc}\left[\frac{(2n+1)L + y}{2\sqrt{Dt}}\right]. \qquad (A.21)$$

Appendix B

Selected diffusion data

$$D = D_0 \exp(-Q/RT), \; R = 8.314 \, \text{J/mol K}$$

B.1   Self-diffusivity of the elements

Matrix	$D_0(\text{m}^2/\text{s})$	$Q(\text{kJ/mol})$
Ag	1.06×10^{-4}	186
Al	1.71×10^{-4}	142
Cr	2.00×10^{-5}	308
Cu	2.32×10^{-4}	206
α-Fe	2.01×10^{-4}	240
γ-Fe	2.20×10^{-5}	268
Ni	2.22×10^{-4}	289
Si	0.9	495

B.2   Solute diffusivity in compounds

Matrix	Diffusing species	$D_0(\text{m}^2/\text{s})$	$Q(\text{kJ/mol})$	Comments
Al_2O_3	Al	2.80×10^{-3}	477	
Cr_2O_3	Cr	0.4	418	
CoO	Co	2.15×10^{-7}	144	
Fe_2O_3	Fe	1.30×10^{2}	419	
MgO	Mg	2.49×10^{-5}	330	
NiO	Ni	4.80×10^{-8}	202	
Ni_3Al	Ni	1.90×10^{-6}	284	

B.3   **Solute diffusivity in alloys**

Matrix	Diffusing species	$D_0(\text{m}^2/\text{s})$	$Q(\text{kJ/mol})$	Comments
Al	Cu	1.50×10^{-5}	126	
	Fe	4.10×10^{-13}	58	
	Ni	2.90×10^{-12}	66	
	Si	9.0×10^{-5}	176	
	Zn	1.40×10^{-4}	129	
Cu	Co	1.93×10^{-4}	226	
	H	1.96×10^{-7}	28.8	
	Ni	1.93×10^{-4}	232	
	Sb	3.40×10^{-5}	128	
	Zn	3.40×10^{-5}	191	
α-Fe	C	2.20×10^{-4}	122	
	Cr	2.53×10^{-4}	240	
	Cu	4.70×10^{-5}	244	
	H	2.30×10^{-9}	6.6	
	Mn	4.9×10^{-5}	276	
	N	8×10^{-7}	76	
	Ni	9.90×10^{-4}	259	
	W	6.90×10^{-3}	265	
γ-Fe	C	1.50×10^{-5}	142	
	Cr	2.2×10^{-5}	268	
	Cu	3.00×10^{-4}	255	
	Ni	7.70×10^{-5}	280	
Nb	H	5×10^{-10}	10.2	
Ni	Au	2.75×10^{-7}	196	
	C	1.20×10^{-5}	142	
	Cr	1.10×10^{-4}	272	
	Cu	5.70×10^{-5}	258	
	Fe	8.0×10^{-5}	255	
	H	6.9×10^{-9}	40.3	$(T < 631\,\text{K})$
		4.8×10^{-9}	39.3	$(T > 631\,\text{K})$
	W	2.0×10^{-4}	299	
Si	Sb	1.3×10^{-3}	383	
Ta	H	4.4×10^{-10}	13.5	
W	Mo	3.70×10^{-7}	460	
	N	5.4×10^{-4}	260	
Zn	Cu	2.00×10^{-4}	125	
Zr	H	7.7×10^{-6}	453	

Appendix C

Selected binary and pseudo-binary phase diagrams

Sources:

Metal diagrams

ASM Handbook, Vol. 3, Alloy Phase Diagrams (1992, ASM International, Materials Park, OH 44073-0002, USA)

Ceramic diagrams

Phase Diagrams for Ceramics Volume I: Basic Volume (MgO–Fe$_2$O$_3$ and MgO–Cr$_2$O$_3$)
Phase Equilibria Diagrams Volume XI (SiO$_2$–Al$_2$O$_3$)
Phase Equilibria Diagrams Volume XII (CaO–ZrO$_2$)

Reprinted with the permission of the American Ceramic Society, P.O. Box 6136, Westerville, OH 43086-6136, USA.

Index to diagrams

Al-Cu

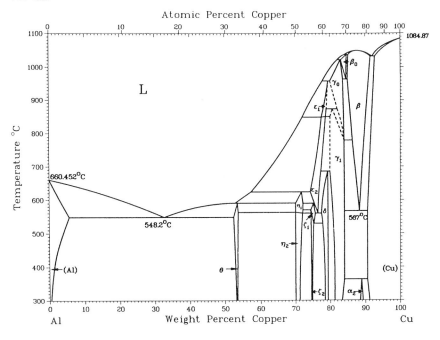

Al-Ni

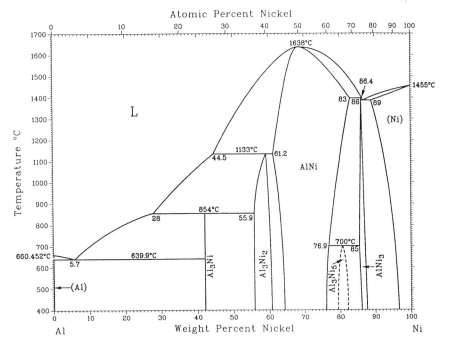

Al-Si

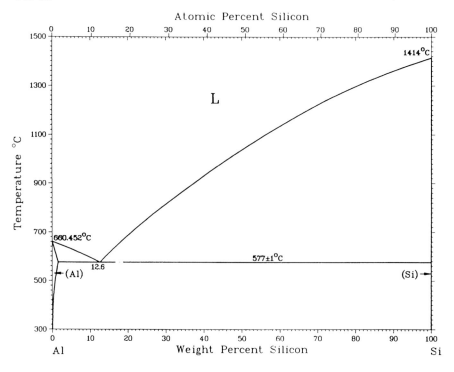

Al-Zn

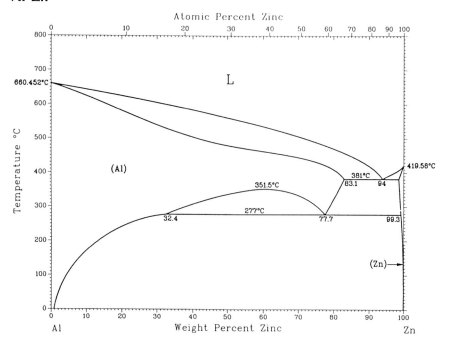

C-Fe

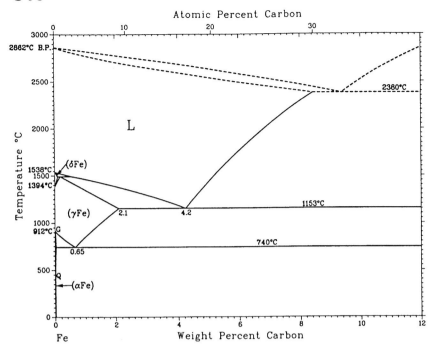

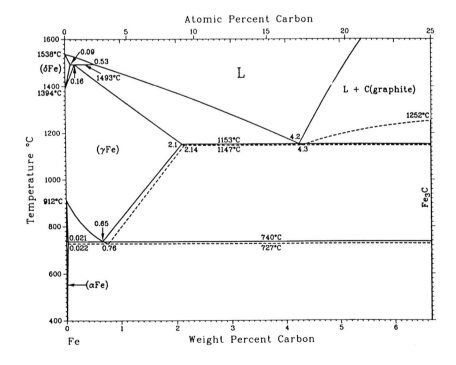

C-Fe (*cont.*)

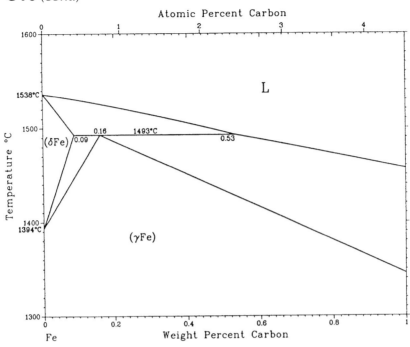

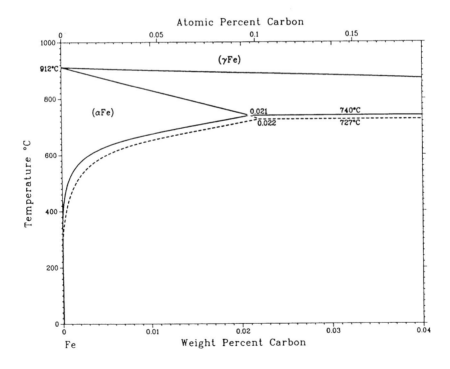

Co-Cu

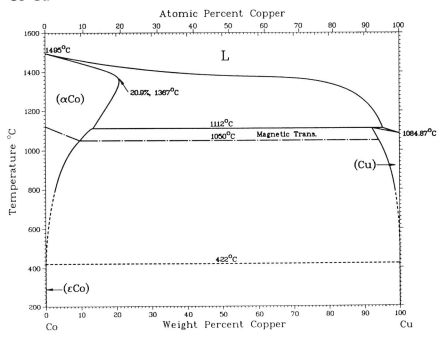

Cr-Fe

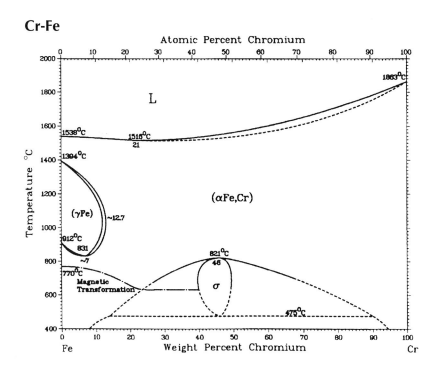

Cu-Fe

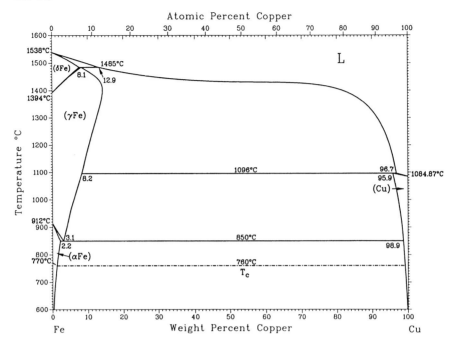

Cu-Mg

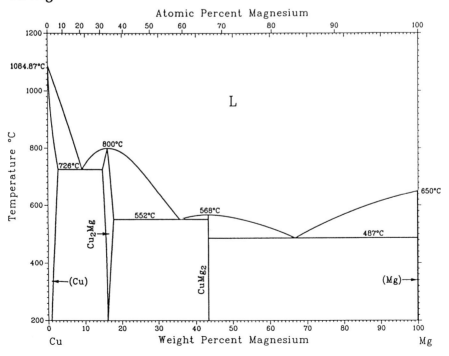

Cu-Nb

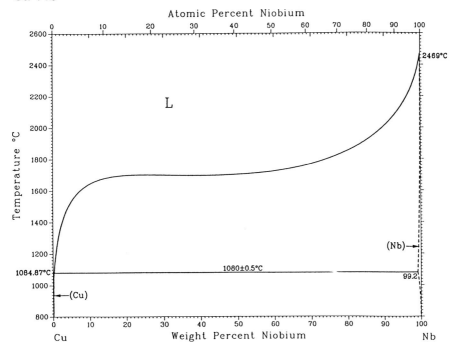

Cu-Ni

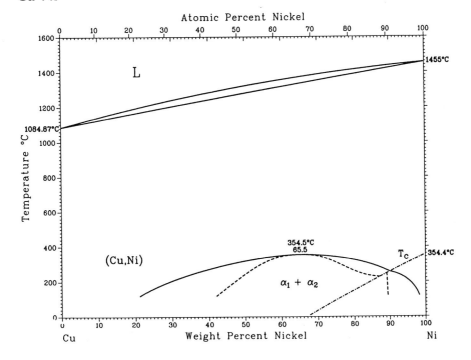

Cu-Sb

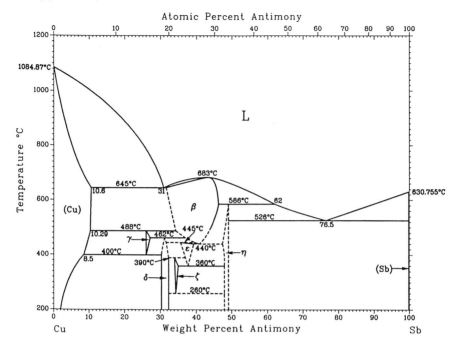

Cu-Zn

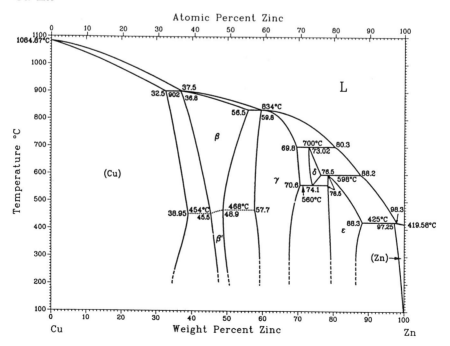

Fe-H

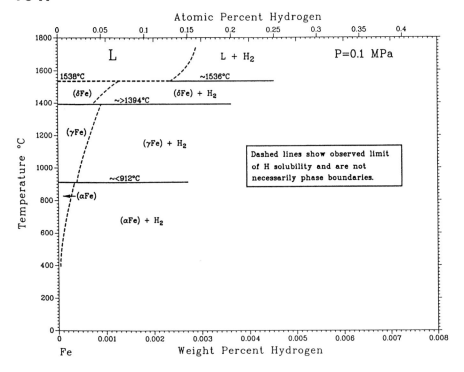

Fe-Mn

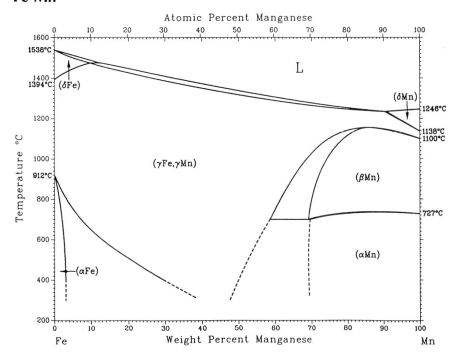

Fe-Zn

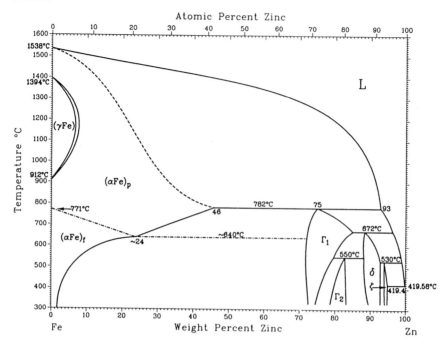

H-Zr

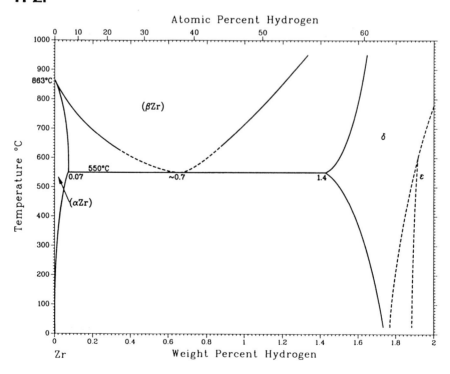

Nb-Sn

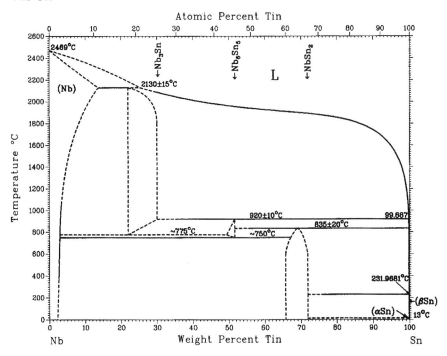

Nb-Zr

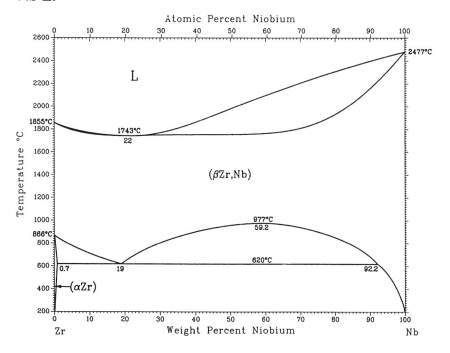

Pb-Sn

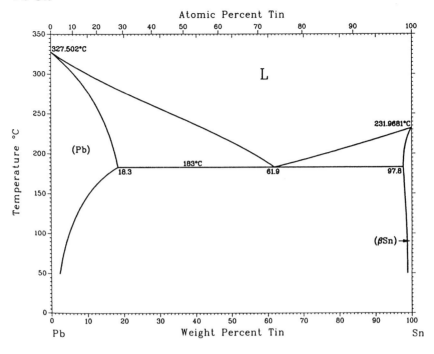

Al$_2$O$_3$–SiO$_2$

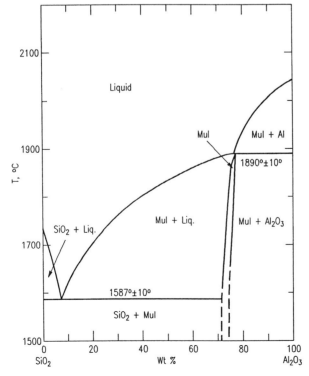

Note: Mul = Mullite (3Al$_2$O$_3$ · 2SiO$_2$)

CaO–ZrO$_2$

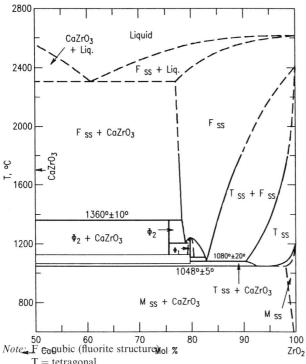

Note: F cubic (fluorite structure)
T = tetragonal
M = monoclinic
Φ_1 = CaO · 4ZrO$_2$
Φ_2 = 6CaO · 19ZrO$_2$

Cr$_2$O$_3$–MgO

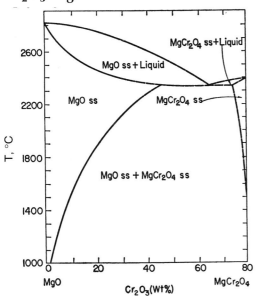

MgO–Fe$_2$O$_3$

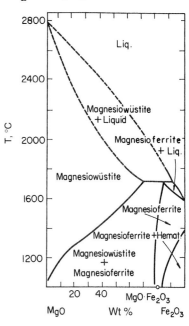

Note: Hemat. = hematite (Fe$_2$O$_3$)

Appendix D

Solving problems by developing conceptual models

D.1 Introduction

Along with learning about specific aspects of matter transport and its relationship to materials engineering, a course in this subject area is an excellent place to learn about the role of modelling in materials science and engineering. Models are used to develop a better understanding of the processes we are trying to control in order to produce better, more reliable and consistent materials. We usually start with a problem (which is often ill-defined, at least initially) and some data. A model is an attempt to draw general conclusions about the behaviour of the system of interest in a quantitative way which allows us to make predictions about the behaviour of the system. Models invariably require certain assumptions and approximations to be made. Thus the model will be limited both in its precision and in its scope or range of applicability. However, a model can be a powerful tool if used carefully, and all materials engineers should have expertise in how to model materials behaviour.

Consider the problem of predicting weather behaviour. A set of observations of many days might lead to the hypothesis that the weather on any given day in a particular location is correlated to the weather on the previous day. Thus if you predict that the weather tomorrow is going to be the same as the weather today you will be correct say 50% of the time. (The exact per cent will of course depend on where you live – this prediction works a lot better in Los Angeles than it does in London). This prediction will be just about as good for

258

the next town 10 km away but will get progressively worse as you attempt to extrapolate to more distant locations. Similarly, your ability to predict weather 2 days or 1 week away will not be as good. Now, until a decade or so ago, this simple model worked about as well as the more sophisticated models used by weather services which (at least where I live) were right only about 50% of the time. More recently, however, much more sophisticated models, involving vast computer power, along with increased availability of data from networked monitoring stations, has greatly increased both the accuracy of weather prediction and the time-scale over which reasonable predictions can be made.

In the current context we are interested in modelling much simpler systems involving mass transport in materials. We generally need to develop a 'conceptual' framework for the system we are dealing with and then develop a (hopefully) simple equation which describes this system. In order to help you gain some facility in doing this I have developed a general method that can be used to approach such problems. This method will enable you to:

- separate out the important features of the problem;
- develop an equation to describe them;
- yield an approximate answer quickly;
- provide a basis for a more accurate answer (following more time and effort) by refining the assumptions made to obtain the quick answer.

Remember that modelling is in part a creative exercise, so no formula will enable you to develop a new model. However, a framework will help to keep you on track.

D.2   General method (to be adapted for particular cases)

A   Identify the problem

What do I know about this situation?
List the things that you know. Above all, draw a schematic diagram of it. This will often indicate other things you know, such as dimensions, and makes it easier to answer the following questions.

What is actually happening in this process?
Determine what is moving or changing, and what is not?

Exactly where on my diagram is the action occurring?
For example, does the process occur at a boundary or within a volume element?

B   *Identify the desired inputs and outputs*

What parameters do I need as inputs?
From the information available identify the variables needed as inputs into a model.

What do I want to know?
What is the parameter I wish to calculate? Can I obtain this information from what I have? Sometimes you need to assume a value.

C   *Identify physical mechanisms*

What controls the rate of action?
In some instances, the process will be dominated by one particular mechanism. Can you identify a likely mechanism? If more than one mechanism is possible you may need to model each of these to see which is rate-controlling.

D   *Target the precision*

How precise an answer do I need?
This depends on how the results are to be applied, so you need to understand the context in which you are developing the model.

How precise a model is justifiable?
This depends largely on the precision of your input data. If some of the important input parameters are known only to within a factor of 2 (say), there is no point in performing a calculation of any greater accuracy.

Can I simplify the geometry?
For example, can one- or two-dimensional cases be considered without losing the essential aspects of the problem? *Be utterly ruthless* about making this assumption; it clarifies the physical situation immensely. *Always* list your assumptions.

Can I make other simplifying assumptions?
For example, it may be possible to assume something is uniform, or if changing, changes linearly rather than in a more complex manner. *Once again, be ruthless.* The trees will remain and the brush goes. *Always* list your assumptions.

E   *Construct the model*

Can I write an equation related to this process?
The previous thoughts and assumptions should have clarified your thinking and helped you identify the key physical processes, so that an appropriate equation can be developed.

Notes on equations

- **Never snatch an equation out of the air.** It must have a *physical* basis which you clearly understand in the context of the problem. Very few equations, such as $F = ma$, are so basic that you can immediately write them down.

- Writing word equations is very helpful to eliminate guessing and omission of important terms. It also clearly separates the various terms.

- Equations often result from something being conserved during a process, for example, total energy, a flux or mass. Ask yourself: **Is anything being conserved in this process?**

- If still no equation has developed, then you need to play with the situation to get a 'feel' for it. Ask questions like: **What would happen if...?** Try taking extreme or limiting cases: for example, what would happen if the temperature changed instantaneously, or what would happen if the volume were very large? If this still does not help you, it is probably time to make additional assumptions in order to simplify your model.

F   Interrogate the model

- Test your equation by using trivial cases (often extreme or limiting cases) for which the answer is obvious.

- Calculate the result according to your equation. The answer will only be as good as the assumptions which you employed.

- Test the model against as much experimental data as you have available.

Now that you understand the physical concepts, you may wish to go back and up-grade the desired precision, by revising the assumptions, and solving the revised equations. This will probably take much longer. In some cases, it will be difficult to solve the more precise problem exactly, but you may be able to determine if the revised assumptions will raise or lower the approximate answer you obtained. Alternatively, you may need to turn to alternative solution techniques, involving computer simulations or finite-element methods, for example.

D.3   **Example**

A set of problems designed to give practice in developing conceptual models appears at the end of this appendix. Let us tackle the second of these, Problem M.2:

Some children are throwing stones at a wire mesh fence. Develop a model to predict the fraction of stones that go through the fence and travel at least 3 metres beyond.

A   *The problem*

What do we know about this situation? We know that a wire mesh fence has a regular arrangement of holes, whose dimensions we can measure. The wire itself also has finite dimensions. The stones are likely to be irregular in shape and variable in size. We will want to develop the model for an 'ideal' stone with a simple shape (*e.g.* sphere) and a well-defined size.

B   *Desired inputs and outputs*

The inputs are the stone radius r, mesh spacing L, and wire diameter w (see Figs. D.1 and D.2). The output parameter is the fraction of stones passing through the mesh.

C   *Physical mechanism*

We will assume that stones which pass *clearly* through the mesh have sufficient velocity to travel at least 3 metres beyond while stones that clip the mesh will either not pass through or will drop a short distance on the other side (*i.e.* will stay within 3 m). We will also assume the stones travel normal to the fence.

For a stone to pass clearly through the mesh its centre of gravity must be at least r away from the wire (see Fig. D.2).

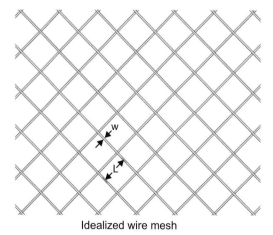

Idealized wire mesh

Figure D.1 An idealized drawing of a wire mesh which offers a conceptual picture of the model.

D   Precision

This is not a very sophisticated problem. However, the precision of our answer will be limited by the assumptions that the stones are spherical and travel normal to the fence. This is likely to introduce errors in the range of 20% or so. Let us initially aim for as simple (*i.e.* low precision) a model as possible. We will therefore make a further assumption that the wire diameter is negligible.

E   Construct the model

The process is well identified in Fig. D.2. The fraction of stones passing through is

$$f = \frac{\text{area of possible mass centres}}{\text{area of the mesh}}.$$

For this simple problem, we can immediately write this down in equation form:

$$f = \frac{(L - 2r)^2}{L^2}.$$

F   Interrogate the model

This is a simple model. Nonetheless we can investigate the result. We see that if the stones are negligibly small ($r \to 0$) all of the stones pass through, while if

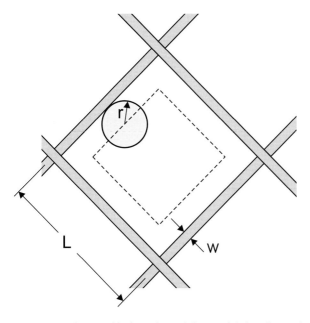

Figure D.2 A magnified version of the model showing a single 'cell' and the limits on the central position a stone must take in order to pass through the fence.

$2r = L$ (stone diameter = mesh size) none pass through. If $2r > L$, obviously stones do not pass through either. So we should refine our result as

$$f = \begin{cases} \dfrac{(L - 2r)^2}{L^2}, & r \leq \dfrac{L}{2} \\ \quad 0 & r \geq \dfrac{L}{2}. \end{cases}$$

We can also consider the effect of relaxing the assumptions we have made. If the wire diameter w is not negligible, the effective mesh dimension is not L, but $L - w$. The result therefore becomes

$$f = \begin{cases} \dfrac{(L - w - 2r)^2}{L^2}, & r \leq \dfrac{L - w}{2} \\ \quad 0, & r \geq \dfrac{L - w}{2}. \end{cases}$$

How big a difference does this make? Well, a typical fence has a mesh of 40 mm and wire diameter of 3 mm. For a stone of 10 mm diameter, $f = 0.56$ if we neglect w, but $f = 0.46$ if we include w, a difference of about 20%. Of course, the difference between the results will depend on stone size. The difference will get larger as $2r$ approaches L.

We could also improve the model by measuring the distribution of stone sizes. From this, we could derive a distribution of fractions passing through the fence in each size range and integrate this to get the overall fraction.

Now that you have seen this process at work, try your hand on the other modelling exercises given below.

D.4   Problems on modelling

Given below are a series of problems to illustrate the modelling approach to problem solving. The object is to determine an approximate mathematical expression to describe each situation. This requires that you make assumptions. Don't be shy about making assumptions, but you should be aware of making them. You can then think about what effect relaxing the assumptions would have on your approximate solution. This may involve developing a more precise quantitative model, or it may simply involve determining what direction the effect would be (*i.e.* will it make the answer larger or smaller).

M.1   Estimate how many ping pong balls would fit into the room in which you sit.

M.2   Some children are throwing stones at a wire mesh fence. Develop a model to predict the fraction of stones that go through the fence and travel at least 3 metres beyond.

M.3   You wish to fill the dog's dish with cold water using your garden hose. The hose has been lying on the lawn connected to the tap but without the water on during a hot day. How many seconds does the water need to run before it gets cold enough to start filling the dog's dish?

M.4   One night you left the hot water tap dripping in your house. You discover it when you get up in the morning. How much did it cost you? (In this case you need to develop not only a simple model, but also think about what measurements you need to make to get input into the model.)

M.5   As you approach a road intersection you notice a truck travelling through the intersection at right angles to your direction of travel. Unbeknownst to you, however, the truck is trailing a long piece of rope. As you drive through the intersection your car drives over the rope. What will happen, *i.e.* what are the conditions that determine whether passing over the rope could affect your car?

M.6   Snow is falling at a uniform rate and accumulating on the windshield of your car.
   (a) Calculate the thickness of snow on the windshield of a stationary car after 1 hour. Thickness is defined in terms of the distance perpendicular to the surface of the windshield.
   (b) Calculate the thickness on a moving car, assuming that no snow is removed from the windshield.

M.7   (a) A knitting needle is poked through a piece of cherry cake several times. It hits a cherry 20% of the time. How many cherries are in the cake?
   (b) Now suppose that you had hit a cherry 80% of the time. Would the assumptions made in the analysis above still be valid? How might the analysis (or the measurements) be improved?

M.8   You are given a two-phase material which contains isolated particles of β-phase in a matrix of α-phase. (*Note:* this is itself an assumption, as it is very difficult to determine whether the particles are isolated or interconnected). You wish to determine the volume fraction of the β-phase and the average size of the particles. Because you cannot poke a knitting needle through solid aluminum, you need to use some other technique, such as optical metallography, to do this. How would you interpret the data you obtain from metallographic sections in order to obtain the necessary parameters?

M.9   Polar glaciers densify by a solid-state sintering process. Snow is deposited continuously at a more or less constant rate (let us say the rate of deposition is M kg/m^2 year). The density of snow when it is deposited is about 30%. Densification occurs due to the pressure exerted by the load of snow. We will use a constitutive law for densification of the form

$$\mathrm{d}D/\mathrm{d}t = k(l - D)p,$$

where D is the relative density (*i.e.* $D = 1$ represents full density), p is the snow pressure, and k is a rate constant. Develop a simple model for the densification behaviour and use it to determine a relationship for the density of a layer of snow, deposited at time $t = 0$, at any time in the future. (For a more complete analysis of this problem see D. S. Wilkinson, 'A pressure sintering model for densification of polar firn and glacier ice', *J. Glaciol.*, **34**, 40–5, 1988.)

M.10   (a) Metallic glass is formed as a ribbon by pouring a stream of molten metal from a crucible onto a rapidly rotating cylinder. The metal solidifies, sticks to the wheel, then is flung off (after a fraction of a revolution). In this way, a continuous ribbon of solid can be produced. Develop an expression for the thickness of a metallic glass ribbon in terms of easily measurable parameters (such as the physical dimensions of various components, the speed of rotation of the cylinder, the depth of liquid in the crucible).

(b) The rate of fluid flow may change with time as the level of liquid in the crucible changes. Indicate how a microprocessor-controlled feedback system might be used to ensure a constant film thickness is produced, as the fluid level fluctuates.

Appendix E

Useful fundamental constants and conversions

Boltzmann's constant	$k = 1.38 \times 10^{-23} \, \text{J/K}$
Avogadro's number	$N_0 = 6.02 \times 10^{23}/\text{mol}$
Universal gas constant	$R = N_0 k = 8.314 \, \text{J/mol K}$
Gravitational acceleration	$g = 9.81 \, \text{m}^2/\text{s}$
Plank's constant	$h = 6.63 \; 10^{-34} \, \text{J s}$

$1 \, \text{eV} = 1.60 \times 10^{-19} \, \text{J}$

$1 \, \text{eV/atom} = 96.5 \, \text{kJ/mol}$

$1 \, \text{calorie} = 4.18 \, \text{J}$

Index